GLOBAL CLASSIC LANDSCAPE DESIGN EXPLORATION HIGHLIGHTS

Global Classic Landscape Design Exploration Highlights

全球经典景观设计探索集锦 IV

《景观设计》杂志社 编

大连理工大学出版社

图书在版编目（CIP）数据

全球经典景观设计探索集锦：全4册 /《景观设计》
杂志社编. -- 大连：大连理工大学出版社, 2011.9

　ISBN 978-7-5611-6520-1

　Ⅰ.①全... Ⅱ.①景... Ⅲ.①景观设计—作品集—世
界—现代 Ⅳ.①TU-856

中国版本图书馆CIP数据核字(2011)第182901号

出版发行：大连理工大学出版社
　　　　　（地址：大连市软件园路80号 邮编：116023）
印　　刷：利丰雅高印刷（深圳）有限公司
幅面尺寸：245mm×245mm
印　　张：60
字　　数：1300千字
出版时间：2011年9月第1版
印刷时间：2011年9月第1次印刷
策划编辑：苗慧珠
责任编辑：刘晓晶
责任校对：万莉立
版式设计：王　江　赵安康　张建实

ISBN 978-7-5611-6520-1
定　价：880.00元（全4册）

电　话：0411-84708842
传　真：0411-84701466
邮　购：0411-84708943
E-mail:dutp@dutp.cn
http://www.landscapedesign.net.cn

目录 Contents

医疗中心 _ **Medical Center**

农业景观 _ **Agriculture Landscape**

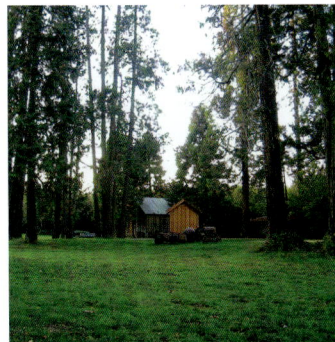

居住区 _ **Residential Area**

目录 Contents

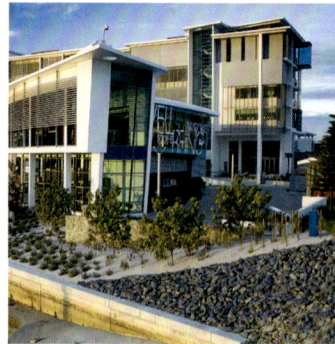

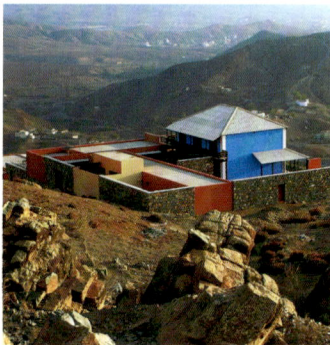

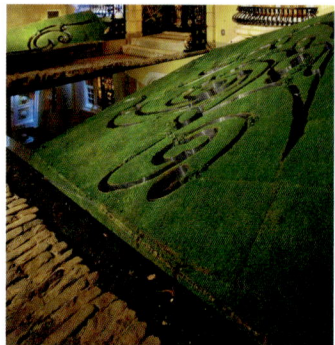

医疗中心

Medical

Center

1

万豪沙漠温泉疗养中心
JW Marriott Desert Springs Spa

撰文：EPTDESIGN　　图片提供：Jack Coyier　　翻译：刘建明

　　该项目于 1987 年开始对外开放，改建后占地面积达 3530 ㎡（由 WAT&G 负责设计），成为南加州规模最大的温泉疗养中心。疗养中心选址于风景秀丽的圣罗莎山麓，标志此处景观营造的新方向。占地约 182 万平方米的全新的温泉景观驱走了热带的灼热，为宾客们营造出另一种景观体验，而这样的设计灵感正是源自沙漠地表的崎岖美。

　　从酒店通往疗养中心的道路被设计成蜿蜒曲折的形状，宾客们在不经意间就能走出酒店、步入休闲的空间。疗养中心的外围，即停车场附近的大部分区域都设有经过雕刻加工的陶制品和有策略栽种的树木。完全依赖灌溉的草坪被除去，代之以精心挑选的沙漠植物，充分展示出岁月更替的自然美和归属于特定季节的色彩魅力，同时也为艺术化的雕刻土墩腾出了展示的空间。堆叠的石头与周围的山峰、金属包裹的建筑立柱和围墙相互呼应。

　　温泉休息室的入口处摆放着一面圆形的石鼓。外面的迎宾广场也呈圆形，广场上引人注目的精美雕刻是由著名设计师 Llisa Demetrios 设计的。巨石、沙漠板岩铺筑材料和起遮蔽作用的木质格架给人以浓浓的暖意。

　　此外，疗养中心还有一些附属设施，包括一个泳池和隐秘的情侣温泉浴区。所有的设计都充分利用了此处景观的简约和沙漠的宁静，并彰显了疗养中心所倡导的放松体验。

Originally opened in 1987, the newly rebuilt 38,000 square foot facility (designed by WAT&G) is the largest spa in Southern California. The setting, in the shadow of the magnificent Santa Rosa mountains, informs the new direction of the landscape. In lieu of the tropical influences which pervade the 450 acre property, the new spa landscape creates an alternate experience for guests and derives inspiration from the rugged beauty of the surrounding desert floor.

The approach to the building from the hotel is arranged as a meandering journey. Guests are subtly invited to leave the scale and styling of the hotel as they prepare to enter a time of rest. The outside world, largely nearby parking lots, was screened through the creation of sculptural earthen mounds and strategically placed trees. Irrigated turf was removed in favor of desert plants chosen for year round beauty, sculptural interest, and seasonal color. Dry-stacked stone, in rich tones emulates the nearby mountains, clad building columns and site walls.

A circular stone "drum" marks the entrance to the spa lobby. The arrival plaza, also circular in form, finds focus on an elegant sculpture grouping by Llisa Demetrios, granddaughter of famed designers Ray and Charles Eames. The warmth of the bronze responds to components of the space such as boulders, desert slate pavers, and sheltering wood trellises.

Additional spaces include a pool area and a private couple's retreat, each designed to capitalize on the simplicity of the view, the tranquility of the desert, and the respite promised in the spa experience.

1

洛杉矶县与南加州大学医疗中心

LAC + USC Medical Center

撰文：Rios Clementi Hale 工作室　　图片提供：Tom Bonner　　翻译：王玲

Rios Clementi Hale 工作室大胆采用取自历史资料并且符合地质特征的大型图案，使之成为洛杉矶县与南加州大学校园概念设计的基础。这些抽象的图案仿佛墨西哥挂毯一样铺在地面上，同时也为路面铺装提供了解决方案并具有场地特色。在场地海拔最低处，抽象的波浪形图案使混凝土铺装充满生机与活力；在场地海拔最高处，俊秀的加那利群岛松使人联想到洛杉矶山区的景象。

该项目占地超过 4 万平方米，拥有 30 米的高程变化，是一处多功能的综合型医院。

设计包括人车交通动线、入口广场、步行广场和花园以及硬质景观与植被景观的整体规划布局。通过对新建筑设计元素的重新诠释——强调建筑与景观之间的联系，并利用场地现有的布局设计，体现出了医院与周边社区的互动关系。

该项目采用大量的元素，并以不同的方式融合在一起，使场地中每一座花园都各具特色，不仅满足各自独特的功能需求，也体现出建筑与场地的和谐共生。设计包括橡树林、通过带状坡道与步道相连的露天草坪广场以及大量的坡道、楼梯和步行广场；其中，步行广场通过彩色混凝土、低矮的树丛以及植有耐旱灌木丛和地被植物的种植区界定出来。

一排由混凝土和金属构成的景观小品是 Rios Clementi Hale 工作室专门为该项目设计的。这些小品由一些重复的组件构成，体现出该项目概念化设计的侧重点。

项目的概念化设计始于对洛杉矶县现有设施及其历史文化和居民的研究。现在的医院保留着许多最初的水磨石地面和一些描述洛杉矶县早期工人阶层家庭和农业根源的瓷砖壁画；进一步研究包括洛杉矶县多样的地质构造，以及从高山到峡谷、再到海洋的地质演变过程。

硬质景观材料

现浇混凝土墙、模压混凝土铺装、喷粉金属场地小品、风化花岗岩

植被

乔木：复羽叶栾树、油橄榄、加那利群岛松、华盛顿葵、加利福尼亚悬铃木、蜜源葵、莓实树

灌木：美人蕉、"塔特蕾"大花假虎刺、木贼、蔓马缨丹、虎刺梅、紫娇花、龙舌兰、迷迭香

1　沿入口车道布置的长凳
2　阶梯状植栽容器使人们
　　有机会触摸植物
3　俯瞰整体布局

Rios Clementi Hale Studios used large bold patterns abstracted from historic references and geological features to form the basis for the conceptual design for the LAC + USC campus. These abstract patterns cover the ground plane like a Mexican tapestry and give way to paving patterns and site features. At the lowest elevation of the site, abstract wave-like patterns animate the concrete paving surface. At the high point of the site, columnar Canary Island pine trees evoke images of the mountain environments of Los Angeles.

The project is the site for the new Los Angeles County + University of Southern California Medical Center (LAC + USC). Extensive site design for the project covers more than 10 acres, manages 100 feet of elevation change and accommodates multiple functions.

Site work consists of the design of automobile and pedestrian circulation systems, entry plazas, pedestrian plazas and gardens, and the overall development and patterning of hardscape and planting elements. By re-interpreting elements of the new building design—relating the architecture and landscape architecture and making use of existing site patterns throughout the entire hospital campus—the design manages the interaction of the hospital with the surrounding community.

The landscape architecture for the facility employs a host of elements that are brought together to create a separate identity for each of the numerous gardens, providing for specific functional needs, while creating an identity for the whole that fits the building to its site. The design includes oak groves, a lawn amphitheater connected by a ribbon-like ramp to a pedestrian walkway above, and an extensive system of ramps, stairs, and pedestrian plazas that are defined by patterning of colored concrete, small tree bosques, and planting areas filled with drought-tolerant shrubs and ground covers.

A line of concrete and metal site furnishings were developed by Rios Clementi Hale Studios expressly for the project. The furnishings are composed of a system of repetitive parts and echo the overriding conceptual design of the project.

The narrative for the conceptual design of the project began with research into the existing LA County facility, and the history of Los Angeles and its inhabitants. The existing hospital retained many of its original WPA terrazzo floors and tile murals depicting working-class families and agrarian roots in early Los Angeles. Further research documented the varied geological formations in Los Angeles, from the mountains to the valley and to the sea.

Hardscape Materials

Cast in place concrete walls; stamped concrete paving; powder coated metal site furnishing; decomposed granite

Plant Materials

Trees: Koelreuteria Bipinnata, Olea Europaea 'Swan Hill', Pinus Canariensis, Washingtonia Robusta, Platanus racemosa, Lagunaria pattersonii, Arbutus unedo

Shrubs: Canna, Carissa 'Tuttlei', Equisetum hyemale, Lantana montevidensis, Euphorbia milii, Tulbaghia violacea, Agave attenuate, Rosmarinus 'Ken Taylor'

温泉理疗中心和酒店

Thermal Bath Therapy and Hotel

撰文 / 图片提供：Jensen & Skodvin Architects　　翻译：王玲

该项目位于奥地利南部的 Bad Gleichenberg 村葱郁山谷中的一个公园保护区内。它包括三部分：一个公共温泉池、一个拥有几间餐厅和咖啡馆的四星级酒店以及一个理疗中心。理疗中心针对不同的理疗按摩和其他医疗服务设置了大约 50 间理疗室。该项目的建筑共分 3 层，总建筑面积为 17 500 ㎡。

项目场地毗邻一座极具当地特色的公园。设计的主要目标之一就是保护那些历经 150 多年洗礼的参天古树，设计师试图在这些古树之间寻找到适合如此大面积的项目的场地（大约 10 000 平方米）。建筑富于纹理，曲线流畅灵活，证明了建造具有丰富几何形状的大体量建筑是可行的。此外，建筑造型也尊重原有树木的位置，与它们相得益彰。

屋顶露台之间通过楼梯相连，使人们能够自由进出不同区域，同时整个建筑也充当了公园的延续——只需在不同的屋顶露台上漫步便可实现在公园漫步的想法。参天古树遮蔽着露台，人们站在露台上就仿佛置身于树顶一般。室外泳池毗邻一些古树，使人们产

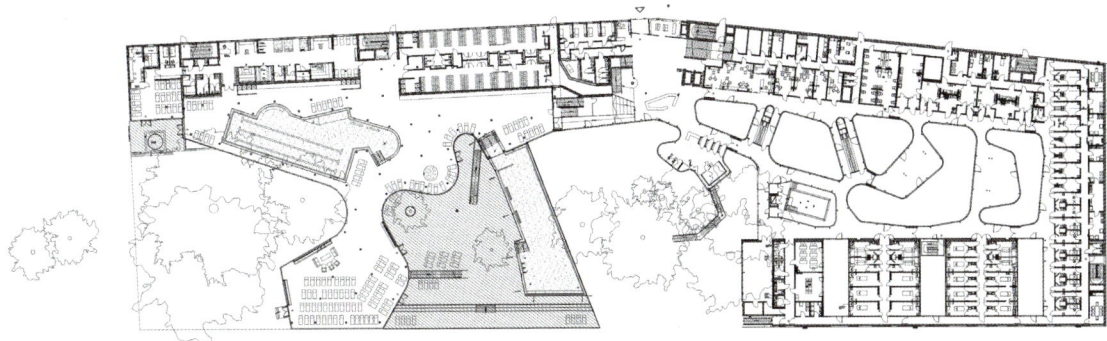

生似乎徜徉在公园的美丽湖泊之中而非泳池中的错觉。

酒店立面上覆盖着天然的落叶松木，随着岁月的流逝这些落叶松木会慢慢地与周围景观的颜色相似。这一设计使大体量建筑与原有公园的环境完美地融合，并注重对自然原生态的保护。

享受全套的理疗服务需要数天，有许多不同的理疗方法，如在小型私人理疗室中的各种按摩和沐浴、将温度降至零下110°C的冷冻疗法等。不同的理疗室之间是通透的等候区，而公园就近在咫尺；等候区环绕着院子，拥有充足的采光条件和良好的景观视野，人们仿佛置身于花园中一般。

温泉理疗的概念非常先进，它利用整体医疗（holistic medicine）的最新见解，并基于一种理念——即客人自己才是最了解自己身体不适和需求的专家。作为理疗的一个部分，这里提供热量只有2720J的4道菜作为晚餐。最近，这里的餐馆被评为奥地利最佳餐馆之一。

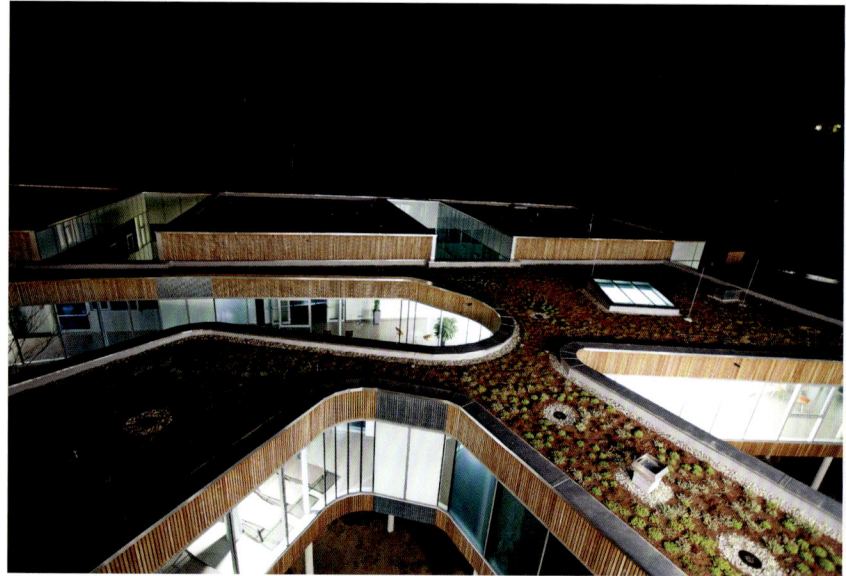

The project is situated in a protected park in the middle of a lush valley in the village of Bad Gleichenberg in the south of Austria. The project consists of three parts: a public thermal bath, a four star hotel with several restaurants and cafes, and a treatment area with about 50 different rooms for several different therapeutic massages and other types of medical treatments. In all the house is about $17,500m^2$ in three levels.

The building site is at the edge of an existing park which is considered a major quality of the area. One of the main objectives has been to preserve a number of more than 150-year-old huge trees. We decided to try to fit the relatively large footprint of the house (about $10,000m^2$) in between the grand old trees. The program of the building is quite fine grained and flexible and it turned out to be possible to accomplish this type of geometric adjustment of the large volume. The building has for large parts been given its shape as the result of this polite way of placing it among the existing trees.

We made terraces on the roofs with stairs between so

that the guests can move in and out among different areas of the house, which makes the whole building functions as a continuation of the park itself. It is possible to experience a walk in the park by walking just on different roofs of the house. Several very large trees hang over the terraces so that one gets the feeling of being up in the tree tops. The outdoor pool is placed close to several old trees and this creates an atmosphere of being in a beautiful lake in a park rather than in a clinical swimming pool.

The hotel is mostly clad with untreated larch that will slowly patinate and get the colors of the surrounding landscape floor. This will add to the feeling that this quite large building is respectfully placed in the park, with the intention not to disturb the existing nature and qualities.

A full treatment for the people using the therapeutic area might last for several days and can consist of a number of different treatments, like different types of massages and baths in smaller private treatment rooms, a visit to a cold room with minus 110 degrees Celsius etc. Between these treatments the patients wait in the transparent waiting areas where the park is always close. These waiting areas are shaped around courtyards allowing sun and views to trees, giving the patients the impression of waiting in the park itself.

The treatment concept of the spa is very advanced and uses the last insights in holistic medicine and is also based upon the belief that the guests are experts on their own bodies, afflictions and individual needs. As a part of the treatment the spa offers a gourmet four course dinners with only 650 calories. The kitchen has recently won an award for one of the best Austrian restaurants.

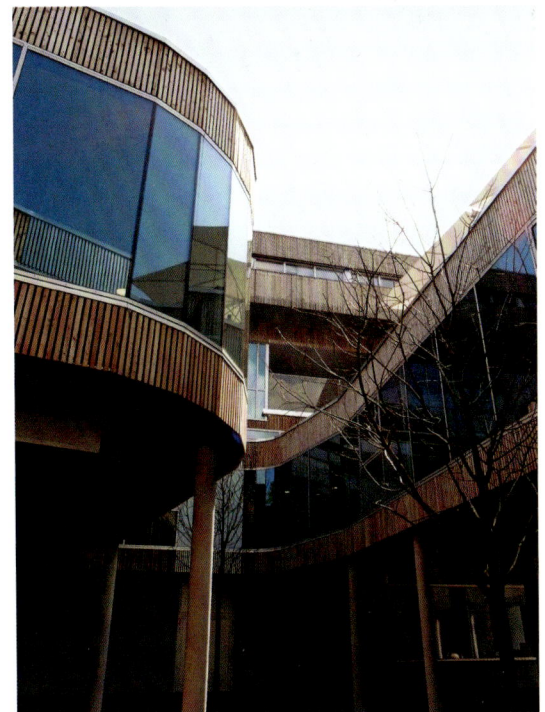

慕尼黑工业大学附属医院

Hospital Rechts der Isar, Munich

撰文：Rainer Schmidt Landschafts Architekten　　图片提供：Raffaella Sirtoli　　翻译：申为军

慕尼黑工业大学附属医院坐落于伊萨尔河畔右侧，拥有一支立志献身医学、技艺精湛的高水平医疗团队，包括医生、护士、科研人员以及技术助理。它是慕尼黑工业大学的一所附属医院，员工总数超过 3700 名，是一所颇负盛名的医疗中心、医学研究中心和医学教研中心。该项目由 30 余个独立的门诊和科室组成，拥有 1100 个床位，每年收治大约 4 万名住院患者和 17 万名门诊患者。

慕尼黑工业大学附属医院十分重视向患者提供全方位的医疗护理。为此，医院成立了 6 个跨学科的临床中心。研究中心紧邻伊萨尔河，其正面是一个英式花园。花园和植物有益于患者康复的理念已经有 1000 余年的历史，在亚欧文化中尤其显著。例如，在中世纪的欧洲，修道院常常会建造优美的花园，抚慰患者并为其带来快乐。在 19 世纪，欧美的大部分医院都拥有花园和植物，这成为了当时医院的显著特征。但在 20 世纪早期，医疗花园已不再盛行了。然而，随着医疗科学领域的迅猛发展，医院管理者和建筑师们开始致力于完善医疗建筑的建设，从而降低感染的风险，同时也为高新医疗技术在功能和效率方面提供完善的环境。这种强调减少传染、功能效率优先的设计理念，影响着国际上数百所知名医院的建设和发展。如今，人们认为这种设计体现为刻板的功能体制，而且给患者及其家属、甚至医院员工的情感方面带来压抑和不适。尽管疾病、疼痛和痛苦难忘的医院经历常常给患者带来高度压力，但医院方面却很少关注医疗环境的建设，尤其是以此来抚慰患者和满足他们的情感需求。近年来，人们逐渐认识到，医疗机构需要一个既高效、卫生，同时也能缓解压力、令人愉悦的环境。达成这种认识的一个重要的推动力是心理健康方面的医学进步。大量研究证实，压力和社会心理因素可大大影响患者的健康状况。这一发现意味着，医院在进行设计时除了需要关注减少感染和注重效率的因素外，患者的情感需求也应被优先考虑。医疗科研人员提出，有助于舒缓压力的健康环境必须在建造新型医院设施时被充分考虑，例如令人心情愉悦的环境和社会支持。

关于这一事实的研究还是有限的，但是越来越多的科学证据表明，观赏花园在一定程度上能够缓解患者压力和改善身体状况，这重新勾起了全球医院和医疗机构对营建观赏花园的兴趣。

该项目的景观设计灵感源于伊萨尔河。由高台种植槽构成的此起彼伏的屋顶花园与伊萨尔河如出一辙，在空地上点缀着树木和花草，这种自由布局形式与河边树木数量相呼应，为人们提供了一个散步以及休闲娱乐的良好场所。

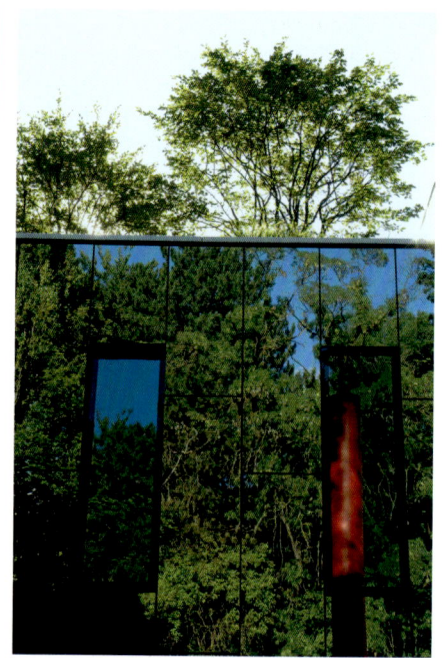

The university hospital rechts der Isar on the right hand side of the river Isar serves Munich and the world with a highly skilled team of dedicated doctors, nurses, research scientists, and technical assistants. It's a university hospital of the Munich Technical University. With a workforce of over 3,700 personnel, the university hospital is a renowned center for the care of the sick, for medical research, and for the teaching of medicine. The rechts der Isar is composed of more than 31 separate clinics and departments treating some 40,000 in-house patients and 170,000 out-patients yearly.

The 1,100-bed hospital rechts der Isar places great emphasis on its ability to give comprehensive care to the individual patient. To achieve this, the clinic has established six interdisciplinary centres. The Klinik is

located directly close to the river Isar and to the English garden. The belief that plants and gardens are beneficial for patients in healthcare environments is more than one thousand years old, and appears prominently in Asian and Western cultures (Ulrich and Parsons, 1992). During the middle Ages in Europe, for example, monasteries created elaborate gardens to bring pleasant, soothing distraction to the ill (Gierlach-Spriggs et al., 1998). European and American hospitals in the 1800s commonly contained gardens and plants as prominent features (Nightingale, 1860). Gardens became less prevalent in hospitals during the early decades of the 1900s, however, as major advances in medical science caused hospital administrators and architects to concentrate on creating healthcare buildings that would reduce infection risk and serve as functionally efficient settings for new medical technology. The strong emphasis on infection reduction, together with the priority given to functional efficiency, shaped the design of hundreds of major hospitals internationally—that are now considered starkly institutional, unacceptably stressful, and unsuited to the emotional needs of patients, their families, and even healthcare staff (Ulrich, 1991; Horsburgh, 1995). Despite the intense stress often caused by illness, pain, and traumatic hospital experiences, little attention was given to creating environments that would calm patients or otherwise address emotional needs (Ulrich, 2001). A growing awareness has developed in recent years in the healthcare community of the need to create functionally efficient and hygienic environments that also have pleasant, stress reducing characteristics. An important impetus for this awareness has been the major progress achieved in mind-body medical science. A substantial body of research has now demonstrated that stress and psychosocial factors can significantly affect patient health outcomes. This knowledge strongly implies that the psychological or emotional needs of patients be given high priority along with traditional concerns, including infection risk exposure and functional efficiency, in governing the design of hospitals (Ulrich, 2001). It also follows that conditions or experiences shown by medical researchers to be stress reducing and healthful, such as pleasant soothing distractions and social support, must become important considerations in creating new healthcare facilities. The fact that there is limited but growing scientific evidence that viewing gardens can measurably reduce patient stress and improve health outcomes has been a key factor in the major resurgence in interest internationally in providing gardens in hospitals and other healthcare facilities.

The landscape design of the hospital rechts der Isar uses an interpretation of the River Isar for the whole garden designs. The roof gardens remain a wavy character to respond to the river. Raised planters modulate those areas. The rest of the open space is characterized by scattered trees and flowering plants. This free layout corresponds to the tree population along the river and invites visitors for a walk and recreational use.

蒙特癌症康复中心

Monter Cancer Center

撰文：Dirtworks PC　　图片提供：Barry Halkin　Dirtworks PC　　翻译：董桂宏

总平面图

每年约有 3000 多位癌症患者慕名前来北海岸大学医院和长岛犹太医疗中心接受治疗。蒙特癌症康复中心附属于高级医药中心，是一家拥有现代化医疗设备且推崇自然康复疗法的癌症门诊医院。

该项目占地面积约 3437 平方米，是由一座历史悠久的仓库改建而成。康复中心的天窗长约 37 米，采光充足，这也是其进行自然康复疗法的一大优势。在康复中心的一期工程中，设计师沿长廊修建了一系列的接待室、候诊室和门诊室，由于长廊宽敞明亮，人们幽默地将其称之为"主街"。

该项目的景观以"竹林"和"桦树林"为主要特色。在康复中心室内，沿"主街"生长着长约 201 米的竹林，其竹叶交错、竹色青葱，静谧的竹林是人们休闲交谈的理想场所。竹林的设计可谓是一箭双雕：一方面，人们在竹林之中可以享受静谧和悠闲；另一方面，医护人员又能够方便地观察到患者的活动情况。

在室内引入竹林可能导致部分癌症患者感染土壤中的病菌，基于这一考虑，设计师请教了很多研究传染性疾病的专家，并且与康复中心的董事会进行了积极协商，最终多方一致决定在竹林生长的土壤上面覆盖一层黑色鹅卵石和一层特殊织物，以保护患者免受土壤中病菌的侵袭。

在室外，笔直的桦树林掩映着康复中心的入口，同时也起到了指引方向的作用。设计师独具匠心，在停车场周围也种植了些许桦树，一棵棵桦树迎风挺立，让患者无时无刻不感受着昂扬向上的积极氛围。

North Shore University Hospital and Long Island Jewish Medical Center provides outpatient cancer services to more than 3,000 people annually. The Monter Cancer Center, part of the Center for Advanced Medicine, represents the Health System's vision to create a new cancer outpatient facility that incorporates the restorative qualities of nature as a complementary healing force.

Located in a converted historic warehouse, the 37,000 square-foot facility uses the building's distinctive 120 foot-long skylights to incorporate natural light into the public and patient care areas, a natural healing advantage. As the first phase of a new comprehensive care center, the design incorporates a series of atria, reception and waiting areas along a spacious corridor called "Main Street".

The landscape concept features bamboo and birch "groves". Inside, bamboo planted along the 1/8-mile-long "Main Street" connects patients to nature. Staggered bands of bamboo create lush, serene settings for relaxing and quiet conversation. The bamboo trunks provide a sense of privacy from passers-by while allowing staff to monitor the various seating areas.

The design addresses the health concern that overexposure to bacteria carried in plants and soil can be harmful to some patients undergoing treatment. Working closely with the Center's Director and Infectious Disease Control Specialist, the plants are submerged in protective strata of black river pebbles and landscape fabric. The handsome natural stone mulch, flushing with the surrounding pavement, shields patients from the contaminants in the soil.

Outside, groves of birch trees define the major points of entrance, which assist in visitor orientation. The birch trees also strategically screen nearby traffic while framing views of the surrounding landscape.

1 室内采光充足、植物茁壮生长

2 "主街"与"竹林"

3 静谧的竹林

英飞凌新总部

Infineon's New Head Office

撰文：Rainer Schmidt Landschafts Architekten　　　图片提供：Raffaella Sirtoli　　　翻译：王玲

该项目的设计与英飞凌公司新总部大楼的名称"CAMPEON"可谓是相得益彰。"CAMPUS"（校园）和"INFINEON"（英飞凌）组成的新词"CAMPEON"将充满创造力与活力的校园与世界知名半导体制造商英飞凌公司的企业精神完美地结合在一起。

从建筑的角度来说，校园首先要具备进行知识交流的必要的基础设施——研究和学习空间、交流空间以及运动休闲空间，无论这些空间是室内的还是室外的。六个被称为模块的建筑体被细分为小单元的建筑体——内部设有管理部门、研究部门、开发部门、行政部门的办公室和实验室。这些模块建筑仿佛身居花园之中，与环绕的清流和如洗的碧空交相辉映。

校园中设置了餐厅、咖啡馆、酒店、购物商场和健身中心。此外，还将为有子女的员工建设一座幼儿园。露天网球场、足球场、沙滩排球场、慢跑和健身跑道也都一应俱全。"CAMPEON"的独特之处在于这里的景观、商店和咖啡馆面向所有人开放，当地居民也可以像英飞凌公司的员工一样在这里悠闲地散步、购物。

校园被细分为实际办公区和自然主义风格的人文公园。人文公园与原有的 UNTERHACHINGER 运动公园浑然一体，形成一个面向慕尼黑南部公众的连贯的公园景观。土堤不仅将景观与高速公路分隔开来，还充当了有效的隔音屏障；它仿佛龙尾一般蜿蜒盘旋在雕塑般的土块上，似乎是阿尔卑斯山脚下冰碛景观的依稀再现。马蹄形池塘环绕着 CAMPEON 的模块建筑，周围是滨水步道和横跨池塘的人行小桥。就像心脏的

两个心房一样，办公区也被分为两个区域，每一分区都拥有模块建筑和东西向的间隙空间；在两个分区之间是一条宽敞的、绿毯般的南北向草坪。

人文公园

L形的人文公园环绕着西部和北部的模块式景观。景观设计师在公园的边缘种植橡树、白蜡树和桦树，中间则是一块不规则的几何状草坪。水结碎石路面贯穿整个公园——向南通向运动场，向东与人行桥相连，并一直延伸到庭院。北面L形人文公园较短的区域将建设一座临时性公园，未来将沿着北面的道路扩建至庭院。

池塘和滨水步道

无论是视觉上还是生态上，水都能够营造出一种独特的氛围。68 000 ㎡的环形池塘不仅是员工工作之余的休闲场所，而且还能有效地改善小气候，充当雨水蓄存池和渗透池。池塘水是由雨水和融化的雪水供

给，为了提高水质，池塘里的水每年都要彻底更换一次。除了每年5月~9月收集到的雨水以外，规划人员还引入了少量的地下水，并将池塘水通过芦苇过滤槽这种"天然净化厂"进行净化。面向公园和土堤隔音屏的地方，规划人员设计了一个适于芦苇和水生植物生长的平坦的湖岸；面向庭院的地方是植有树木的滨水步道和广场。在这里，湖岸要么是绿草茵茵的陡坡，要么是伸入水面的混凝土边缘；小广场在模块建筑的一端伸入水面，圆锥形的白柳掩映其间；中间的下沉花园里种植着婀娜的垂柳和摇曳的小白杨。

绿色空间

宽敞的草坪勾绘出校园绿色空间中心地带的主要特征，精美笔直的碎石小径纵横于草坪之间，楔形的造型使草坪的地形彰显出些许灵动之意。挺拔的松树在绿色空间的南北两侧形成轻盈的华盖——一些松树群上长满藤蔓玫瑰；黄瓶子草和玉兰树点缀其间，使松树群显得更加野趣横生。沿建筑铺设的路面是混凝土路面、地砖、草坪砖和水结碎石路，与它们相连的则是6m宽的主路。

模块建筑以及内庭院之间的空间

景观由东向西一直延伸至模块建筑之间，设计师在建筑之间设计了通往每座办公楼的道路。这些道路采用沥青砂胶路面，道路两边是蜿蜒曲折的种满高茎草的种植槽。

沟渠可用于收集过量的雨水，低茎草和帚状的白杨沿着种植槽苗壮成长。一些小巧精致的内庭院隐藏于模块建筑凹凸处，成为远离主路的私密花园。草坪上随意种植的果树被格架框起来，一直延伸到模块建筑之间；在花香四溢的季节里，它们将设有长凳的粗石广场装点得更加舒适宜人。

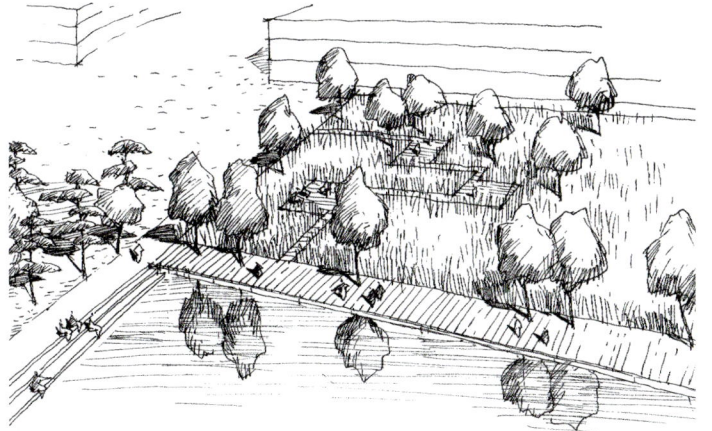

The programme first becomes evident in the name chosen for the new head office: "Campeon". This combination of "campus" and "Infineon" indicates that the creativity and vigour of a university campus combines here with the enterprising spirit of a world-famous semiconductor manufacturer.

Architecturally speaking, a campus above all provides the necessary infrastructure for the effective exchange of knowledge: spaces for research and work as well as spaces for communication and spaces for sports and recreation, whether indoors or outside in the open air. Six of the so-called modules—one to three-storey building complexes subdivided into small units-house—the offices and labs for management, research, development and administration. The modules standing in the park spread out with water surrounding it and sky overhead.

A cafeteria, cafés, a restaurant, a shopping mall and fitness centre also belong to the campus. A day care centre will be set up for the children of the employees. In the open space there will be tennis, soccer and beach volleyball courts, and jogging and fitness trails. Its special feature: the Campeon landscape, shops and cafés are open to everyone. Local residents and neighbours can go for walks and shop here just as Infineon employees do.

Consequently, the campus subdivides into the actual office complex and a naturalistically designed people's park being the direct extension of the existing Unterhachinger Sports Park. This creates one continuous park landscape open to the public in the south of Munich. An earth embankment blocks off the landscape from the motorway and provides a noise barrier. The embankment rises from the plain like a dragon's tail, and if you like you can see the moraine landscape of the foothills of the Alps in its sculpturally formed clods of earth. A U-shaped pond encloses the modular buildings of the Campeon, bordered by a water-side promenade and crossed by footbridges. Like a heart, the office area has two halves, each with modules and interstitial spaces oriented in an east-west direction. In the middle, a spacious green runs through the grounds from north to south.

People's Park

The L-shaped people's park en-closes the modular landscape on the west and north sides. The landscape architects are planting its edges with oak, ash and birch trees, and seeding a lean lawn in the centre. Waterbound path-ways run through the park and lead in the south to the sports grounds and in the east to the foot-bridges and across the pond into the Campeon grounds. The short end of the L in the north will be a park for the time being. Here, building ex-tensions are projected along the north access road to the grounds.

Pond and Waterside Promenade

Water creates atmosphere both visually and ecologically. The ring-shaped, 6.8-hectare Campeon pond not only attracts users during break time on a workday but also improves the microclimate and serves as a retention and seepage basin for the rainwater. The pond is fed by rainwater and snow. In order to keep water quality high, the water is changed

completely once a year. The planners do this by introducing small quantities of groundwater besides the rainwater from May to September and directing the pond's water through reed filtering tanks—forming a natural purification plant—during this time. Towards the park and the noise barrier embankment, the planners gave the pond a flat shore, where reeds and aquatic plants can grow. Towards the Campeon grounds, they set up a waterside promenade and squares with stands of trees. Here the shore drops in the form of a steep grassy bank or a concrete edge down to the water. Like bas tions, small squares push into the pond at the ends of the gaps between the modules. Conical white willows project over them. The sunken gardens placed in between feature weeping willows and trembling poplars.

The Green

Broad lawns characterise the central strip of the campus green with perfectly straight crushed-stone footpaths cutting through them. Rising like wedges, the lawns create a slight movement in the topography. Tall pines form a light canopy on one side of the green in the north and on the other in the south. Some groups of pines have climbing roses growing on them; trumpet and magnolia trees loosen up the strips of pines. Running along the buildings are strips of concrete paving and tiles, lawn tiles and a waterbound surface. Integrated into them is the six metre-wide main access route.

Spaces between the modules, inner courts

Running from east to west, the landscape extends through the Campeon between the modular buildings. The landscape architects also lead the routes to the individual offices through the interstitial spaces. A strip of asphalt mastic identifies the routes; bordering these are troughs with zigzag edges and with tall grasses growing in them.

Ditches for collecting rainwater extend here, and excess rainwater can seep in. Low grasses with fastigiate poplars arising from them grow around the troughs. Hiding in the projections and recesses of the modular architecture are several little inner courts. They serve as secluded gardens for pausing in, points of rest away from the main routes. The informally scattered fruit trees on their lawns provide experiences of fragrances and blossoms. Fruit trees growing on trellises move into the architectural grid and lend the coarse gravel covered squares with their benches a cosy character.

平衡与安宁的绿色设计 —— 十字架上的欧米茄

Green Design for Balance and Peace — Omega at the Crossings

撰文：The Office of James Burnett   图片提供：Matt McKinney The Office of James Burnett   翻译：张晶

该项目位于景色优美的德克萨斯州希尔县，是一家专门提供温泉水疗和保健的中心。其创办者是一对夫妇，丈夫是戴尔公司的一名退休管理人员，妻子是有着 30 年工作经验的精神科医生。他们将各自的专业知识结合起来，试图构建一个生活社区，以"启迪人们寻找自身工作、生活乃至精神世界的平衡与安宁。"该项目共有 8 栋客舍、70 间客房。园内还设有车库、接待处、动物保护区、餐厅和保健中心，其中保健中心可以为团体及个人提供温泉水疗、瑜珈、气功、日本灵气疗法 (reiki)、生活教练以及其他多项保健服务。

为尽量避免干扰场地的生态环境，如濒临灭绝的金颊林莺的栖息地，客户与美国渔业与野生动物局达成了一份关于开发占地面积约为 137 700 平方米的项目的协议。景观设计师与建筑师共同评估了该场地的环境，规划方案最大限度地保留了阔叶树、高大的杜松、天然草地和场地中的巨大岩层。同时，工程承包商也尽可能地减少施工对建筑物周边地区的不良影响。

在该项目的设计与施工过程中，客户最重视的是如何体现绿色设计理念。所有建筑物都应用被动式太阳能设计策略，设备及材料多选用粉煤灰混凝土、木塑装饰板、纤维水泥板和低流量设备。景观设计师在设计中尽量减少使用场地原有的块石路面，从而使雨水尽可能多地渗入爱德华兹蓄水层。园区内铺设了风化花岗岩小路，同时为了减少路面的吸热量，无法渗透雨水的块石路面均选取浅色石材。工程中大量应用德克萨斯石灰岩，而且都购于当地的采石场。设计师将收集到的雨水和园区内的生活污水再次使用到灌溉系统中，但不涵盖接待处附近的两个地方——那里种植了本地的耐旱植物，并且都修剪得非常整齐美观。虽然该项目在施工时并未获得 LEED 认证，但是目前已获得 LEED 黄金级认证。

设计师在主车道通往入口处用德克萨斯石灰岩修砌了长长的围墙，"Crossings"的徽标就镌刻其上。游人驾车驶过大门便可到达车库，那里存放着可供在园区内代步的电动高尔夫球车。园区内的车道狭窄曲折，从入口处到掩映于杜松林中的建筑物大约有 805 米的

总平面图

1　客房门前茂密的植物

2　建筑选址尽可能保留当地原有林木植被

3　场地周围的小场景为人们交谈或沉思提供了较为私密的空间

4　平和园的大门由石灰石板修砌而成

车程。游人到达接待处后，会有专人带其前往他们预订的服务设施。设计师围绕着几处客舍和一个小型会议中心的四周设置了环型车道，并在车道沿线设置了几个停车场。

　　会议中心和主干道之间由风化花岗岩铺成的小路连通。游人可以在会议中心的大厅里参与各种联谊活动，也可以在图书馆里读书或参加活动草坪上的各种户外活动。沿着小路信步南行，不远处就是坐落于斜坡上的餐厅和保健中心，在那里可以远眺风景秀丽的特拉维斯湖（Lake Travis）。露天餐厅的设施选取了德克萨斯当地的石材和简单的柚木桌椅，中间保留了几

棵高大茂盛的橡树，充满浓郁的奥斯汀地方特色，完全可以满足多达 350 人的就餐需求。保健中心的温泉水疗设施的周围是别具特色的德克萨斯当地植物园、小块的活动草坪和水疗池，人们可以一边休闲、一边欣赏一览无余的湖光山色。

　　动物保护区距离入口处有几百米远，是园内惟一一处几条干路支线都可以通达的区域。前往动物保护区的路上会经过"平和园"（Balance Garden），这是一处设有围墙的花园，寓意秩序与混乱。设计师在笔直的拦河坝四周栽种了整齐的墨西哥悬铃木和多年生植物，而种植在通向动物保护区的厚石板路两旁的植

物则呈蜿蜒之势。动物保护区的设计灵感源自"十字架"的字面意思，用厚石板路引领游人来到动物保护区的冥想空间，石板上手工雕刻了象征世界宗教的图案。动物保护区后面隐藏着一小块迷宫式的花园，其间设有迂回的小路可以通往园区主干路。园区主干路总长约 4000 米，沿途尽是当地的特色景观。

Located on 210 acres of fragile land in the Texas Hill Country, the Crossings is a destination spa and wellness center founded by a retired Dell Computer executive and his wife, a psychotherapist with 30 years of experience. Combining their respective expertise, the clients sought to establish a community that would "inspire people to find peace and balance in their professional, personal and spiritual lives". With 70 guest rooms distributed in eight lodges, the Crossings campus features a carriage house, a welcome center, a sanctuary, a dining hall and a wellness center that offers spa services, yoga, qi gong, reiki, life coaching and a variety of wellness-related services for groups and individuals.

To limit the disturbance to the project site, which includes habitat for the endangered Golden Cheeked Warbler, the owners reached an agreement with U.S. Fish and Wildlife Service that would limit development to 34 acres. The landscape architect and architect worked together to evaluate site conditions and the site plan preserves the majority of the hardwood trees, large ash junipers, native grasses and significant rock formations found on the site. Additional care was taken by the contractor to limit the impact of construction to the area immediately surrounding the buildings.

Incorporating green design was a client priority throughout the design and construction process. All buildings feature passive solar strategies, fly ash concrete, wood-plastic decking, fiber-cement siding and low-flow fixtures. The landscape architect sought to minimize pervious paving across the site to maximize the amount of rainfall that percolates through to the Edwards Aquifer. Decomposed granite trails are used throughout the site; where impervious paving is used light colors were selected to reduce the buildup of heat. Texas limestone was used extensively across the project and was procured from local quarries. Rainwater and effluent are both harvested on-site to feed the irrigation system and with the exception of two highly-manicured areas near the visitor center, the site is planted with native and drought tolerant plants. Although LEED certification was not pursued during the construction process, the project is currently seeking a Gold certification through the Existing Buildings program.

A long wall of Texas limestone bearing the Crossings logo marks the entrance from the main roadway. The entry drive leads visitors past a gate and carriage house where electric golf carts used for transportation across the campus are stored. The narrow drive winds through the landscape for a half-mile before buildings begin to appear nestled among the junipers. Guests arrive at the welcome center where they are directed to other facilities based on their reservations. Site parking is distributed in small clusters along a loop road that serves several lodges and a small meeting center.

Decomposed granite paths connect the conference center to the main entry where guests may socialize in the main hall, read in the library or participate in an activity on the function lawn. Trails lead south to the dining hall and wellness center which sit on the slope overlooking Lake Travis. Rooted in the Austin vernacular through the use native Texas stone and simple teak furniture, the dining terrace is built around several existing specimen live oaks and can accommodate up to 350 guests for special events. Organized around the spa, the wellness center features gardens of native Texas plants, a small function lawn and an infinity-edge pool that offers commanding views of the lake below.

Sited several hundred yards into the landscape, the sanctuary is only accessible from several branches of the trail network that crosses the site. The path to the sanctuary passes through the Balance Garden, a pair of walled gardens abstracting order and chaos. Organized around a linear weir, the structured planting of Mexican Sycamores and perennials loses its rigidity once it passes the stone slab path to the sanctuary and becomes a meandering stream. Conceived as the literal manifestation of the Crossing, slab path leads guests into the contemplative space of the sanctuary which is with hand-carved symbols representing the world's religions. Hidden in a small clearing beyond the sanctuary, a garden maze provides an introspective detour before rejoining the 2.5-mile trail network that disappears into the native landscape.

1 露天餐厅的四周由橡树环绕，餐厅的地面铺装取材于场地上原有的石材
2 露天餐厅可以俯瞰特拉维斯湖的景色
3 设计师将当地的传统材料与现代设计观念相结合
4 由风化花岗岩铺设的小径，连结了场地上所有的基础设施并减少了不透水路面的面积
5 场地中种植的德克萨斯本地植物，洼地用来过滤和涵养雨水
6 场地围墙的建设尽量避开已有的树木，从而减少了对场地中植被的破坏

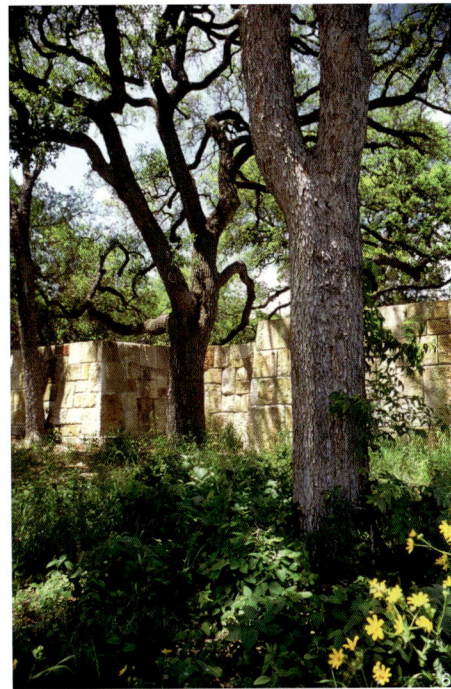

整合自然的内在康复 —— 帕克里克 H.多拉德健康中心

Incorporating Nature's Inherent Resorative Qualities — Patrick H. Dollard Health Center

撰文：大卫·坎普   图片提供：David Alee Dirtworks PC   翻译：张晶

发现中心（Center for Discovery）以其创新的教育、临床和社会经验使伤残人士、病患及其家属能够在这里实现自我价值，享受到更加丰富多彩的生活。该中心位于美国纽约州，总占地面积 140 万平方米，发现中心内的林地、草甸、牧场和农田均展现出该地区特有的乡村田园风光；其设计灵魂就是尊重自然和融入自然。

发现中心的设计人员坚信环境能够影响人们的健康，如果建筑的设计、维修和运营未经检验，势必会带来大量不利的环境问题。设计师将这一理念应用到了发现中心的工程设计、规划和建设之中，并将其体现在帕特里克 H. 多拉德健康中心（Patrick H. Dollard Health Center）的建设中。

尽管该中心地处有着大片农田的偏僻乡村，但这里过去却是废弃的工、农业区，原址上的一些大型建筑极大地影响了水流和水质，甚至干扰了人们的观景视线。设计师在选址时避开了主要的农田，而选择了恢复天然水源流向和当地生态环境这一挑战性极强的

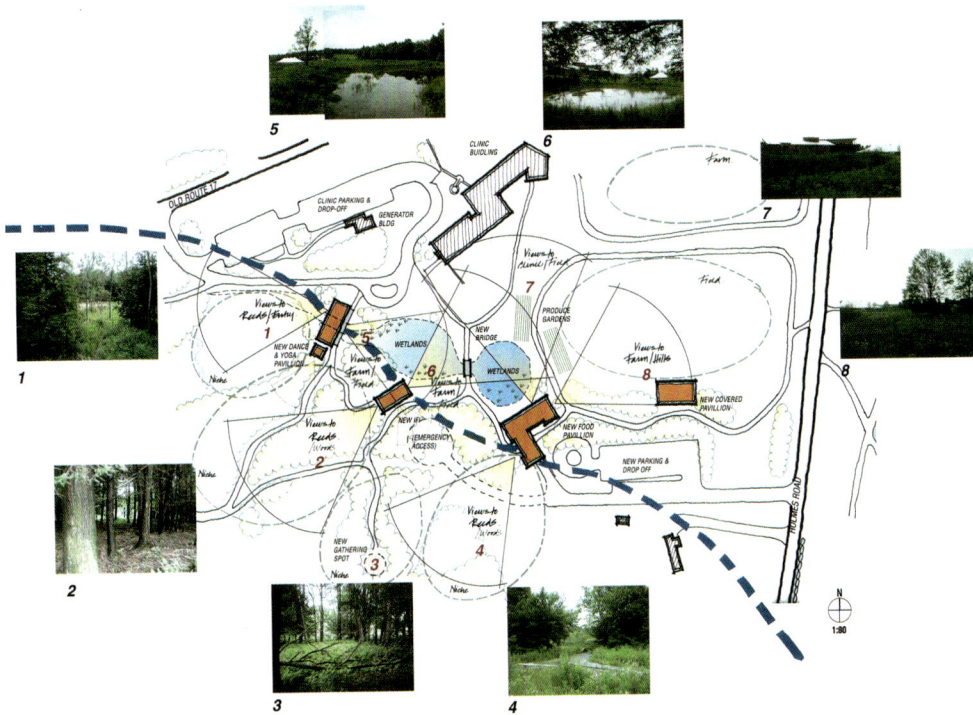

任务。如今，设计师对发现中心园区的内部景观进行了改造和修缮，并在大厦所有内部和室外地面实施了虫害综合防治管理。发现中心终于实现了造景方式、可持续性原则和设计目标的有机统一。

在发现中心园区内的各项工程中，Dirtworks 景观建筑设计公司将可持续性建筑设计及健康理念进一步发扬光大。通过医生、护士、教师和宿舍管理人员的密切配合，设计师精心打造的景观让患者有机会通过康复训练、散步、培育野生及当地植物等户外活动来享受大自然。这些工程包括专门为孤独症儿童开辟的新校园、家庭娱乐中心、舞厅，以及为体质虚弱的成年人新修建的住所。

当地的动植物景观在每一项工程中都是必不可少的一部分，与其交织在一起的果园和灌木篱笆墙既可以划分出入口和行走路线，又可以让人们欣赏到不同角度的多重美景。秀美的景观和广袤的草场令人倍感亲切和喜悦，并且使新的建筑与发现中心大园区整体联通，让人们拥有更多的机会去接触自然、研究自然和探索自然。

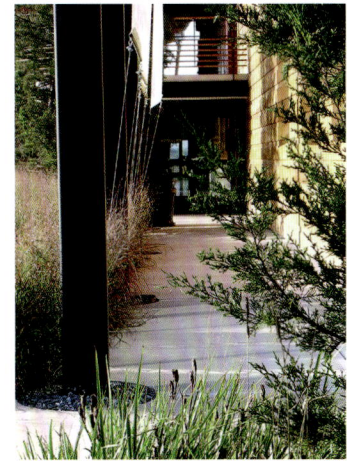

The Center for Discovery offers individuals with multiple disabilities and medical frailties and their families innovative educational, clinical and social experiences designed to enrich their lives through personal accomplishment. Located on three hundred and fifty acres in rural New York State, the facility's setting reflects the area's agricultural and landscape heritage with woodlands, meadows, pastures and farmland. At the heart of the program is a deep respect for and connection with nature.

The Center believes that the health of the environment influences the health of the community and that a building's design, maintenance and operations can have substantial environmental impacts if left unchecked. The Center has transformed the way in which they consider architecture, planning and construction, and health. This concept is reflected in The Center for Discovery's Patrick H. Dollard Health Center, a newly constructed 27,000 sf diagnostic and treatment facility.

While located in a relatively remote rural area with available farmland, the Health Center was developed on a previously abandoned industrial agricultural site, one whose large-scale buildings had substantially impacted water movement and quality, and had blighted the view corridor. In selecting this site, the Center avoided prime agricultural land and began the challenging task of restoring natural water flow and local habitat. Today, the Center for Discovery has reshaped landscaping and maintenance practices across their campus, and launched an Integrated Pest Management program in all buildings and grounds. The Center for Discovery has shifted its landscaping approach into alignment with broader sustainability principles and design goals.

Dirtworks is expanding upon this concept of sustainable building practices and health through involvement in a variety of diverse projects across the Center for Discovery campus. Working closely with clinicians, nurses, teachers and residential staff the landscape design is supporting the Center for Discovery's goal of offering individuals with medical frailties the opportunity to enjoy nature through outdoor activity and education areas, walking trails, wildlife and native plant enhancement programs. The projects include master planning for a new campus for children with autism, a Family Resource Center, Dance Barn, and new housing for frail adults.

In each of these projects, the local vernacular landscape is used as an organizing and unifying element. Orchards and hedgerows weave through the landscape to define entries and circulation and to frame and screen views. Together with sweeping open meadows, the landscape establishes a familiar and welcoming setting, linking the various new buildings to the entire Center for Discovery campus, the region and world beyond, providing rich opportunities to engage oneself with nature, to learn and to explore.

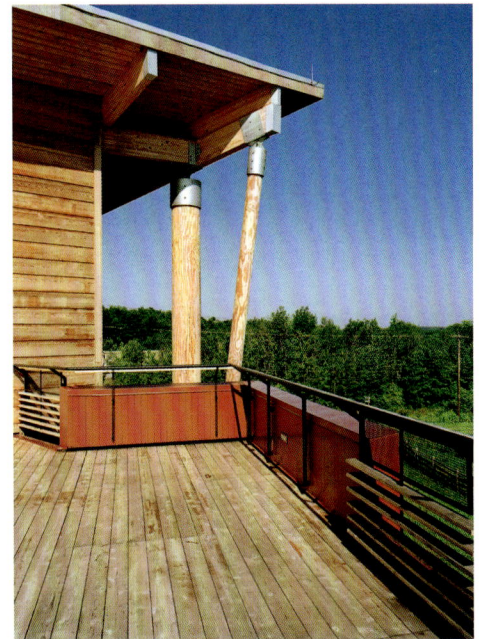

Agriculture

农业景观

Landscape

农业生产景观 —— Tierra Atacama 酒店

Productive Landscape — Tierra Atacama Hotel

撰文：Jimena Martignoni　　　图片提供：Teresa Moller Office　Jimena Martignoni　　　翻译：武秀伟

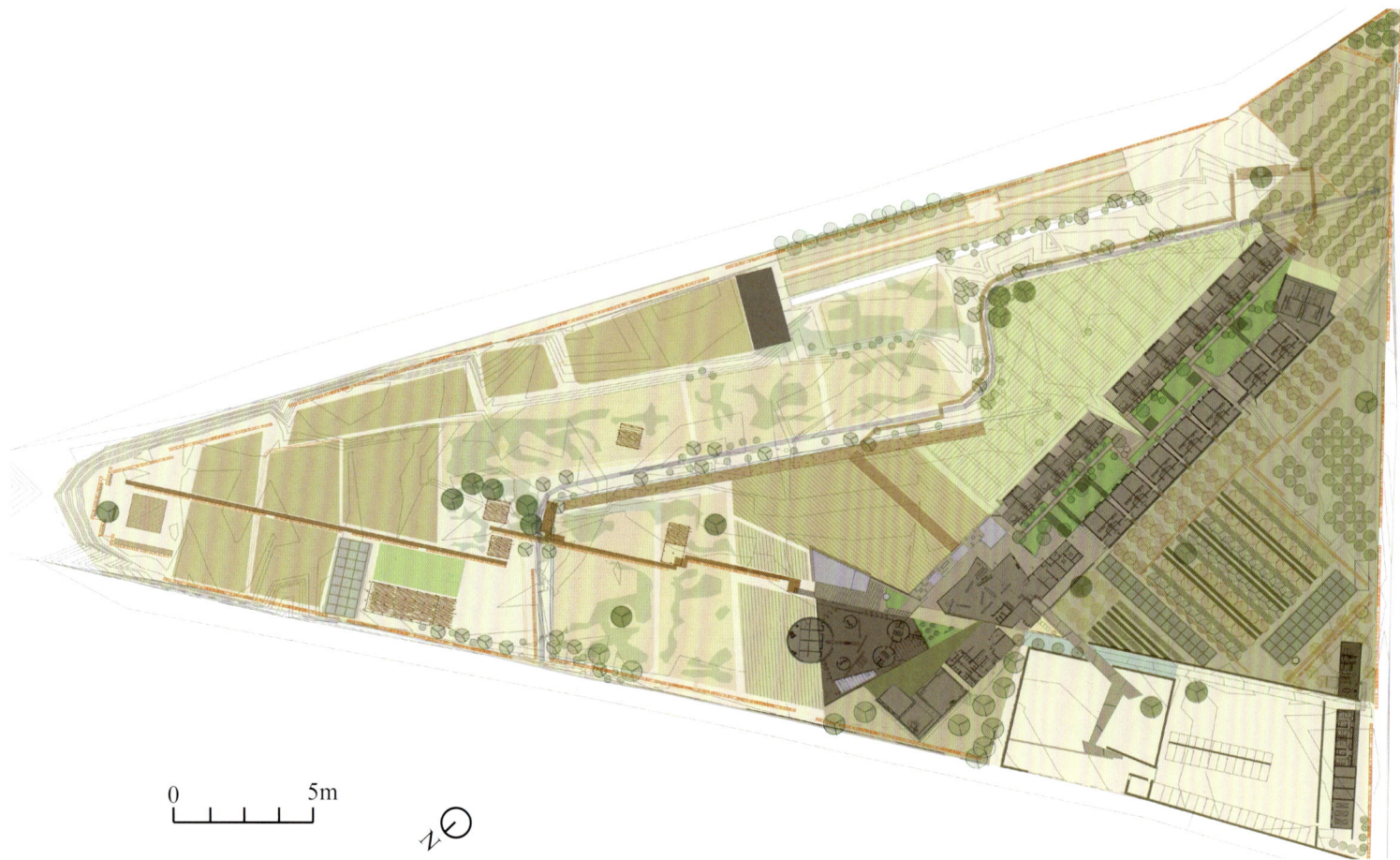

0　　　　　　5m

总平面图

Tierra Atacama 酒店位于智利的圣佩德罗 – 德阿塔卡马小镇，这里是阿塔卡马沙漠中一片天然的绿洲。建造酒店的场地曾经荒废了 30 年，后来被改造为耕地，现在完全呈现出一派农业生产的景观，不仅能够满足酒店自身的食物需求，也可以满足当地居民的食物需求。

酒店周围的水资源丰富。因此，需要利用水资源巧妙地设计入口通道，并在其他方面也要着力体现出艺术性与功能性的完美结合。

Teresa Moller 是该项目的景观设计师，她以诗意的方式来追溯场地景观的起源——这里曾经是耕地，为当地居民提供日常食物，现在要进一步发掘它的潜力。设计师决定依照最初的使用目的对这片土地重新加以利用，使其色调、结构及品味能完美地搭配在一起。这一理念不仅成为建筑设计的典范，也为人与自然的和谐共生树立了榜样。

圣佩德罗 – 德阿塔卡马小镇的特殊气候的形成与安第斯山脉典型的高原冬季降雨有关。亚马逊河大量湿润的气团形成了这种气候，而源于安第斯山脉的圣佩德罗河与维拉马河是该绿洲长期存在的决定性因素。

圣佩德罗河是这里惟一的水源（除地下水）。流经城镇的河水分别被引至不同的区域或部落。这些部落是该地区在哥伦布时代前的社会经济基础，部落里的土地都分配给了个人。如今，这种部落体系依然存在。圣佩德罗镇共有 15 个部落，但当地的大部分居民都已经搬迁到该地区的外围居住，而在中心区域建造了很多旅馆、酒店，以接待大量的游客。

圣佩德罗 – 德阿塔卡马小镇有着一套严格的水资源分配方法和时间表，并由水资源分配管理协会进行监督和管理，这对有效地利用土地至关重要。所有耕地都通过人工修建的主水渠和辅助水渠进行灌溉，灌溉周期为 15 天或 18 天，这种分时段灌溉法是该镇水资源管理系统的基础，可以满足土地的灌溉需求。每公顷土地的灌溉时间规定为两个小时，在实际灌溉中，还要确保水充分渗透土壤。例如，酒店场地不是单纯地用做粮食生产，其灌溉时间取决于其他条件，因此，生产性、社会用途以及场地的使用年数都是灌溉时需要考虑的特殊情况。

该项目占地 55 000 ㎡，有两条辅助水渠贯穿整个场地，而酒店只有权利使用其中的一条。但是这种灌溉方法也有不足之处，比如水流的边界、流速、流量难以控制，容易造成水资源分配不足或严重流失。由于这个原因，该场地不得不采用其他灌溉方法，如滴灌、喷灌、微喷等。这些灌溉方法都可以满足特别的灌溉需求，并能对原有的灌溉方法进行补充，灌溉用水都是从场地上一处 60m 的深井中抽取。

场地中种植的多种农作物反映出安第斯山脉的特色，这些农作物都是经过试种的。玉米、燕麦、大麦和奎奴亚藜（印加地区的典型农作物）都是最能代表当地特色的农作物，向日葵、果树和其他可食用农作物也使当地农作物的种类变得更加丰富。不同的农作物在选择种植位置时要考虑两个因素：首先，要考虑其实用性，即应该满足酒店日常的食物需要；其次，要考虑其对整个场地艺术效果的影响——在靠近酒店设施的地方，场地的色调变化较多；而在离建筑物和庭院较远、土地较多的地方，农作物比较茂盛；场地西部是成片的果园，给面向这边的客房增添了充满活力的视觉体验。

1　木栈道旁的向日葵
2　休息平台

在酒店客房的东面，可以看到远处的高山及火山；大片的波斯菊（一种耐旱植物，开粉色小花）宛如单色的地毯；燕麦与大麦混合种植，交替变换着美景。所有的客房都可以清楚地看到位于东楼与西楼之间的线性空间，设计师在这里设计了一处私密花园，里面种植着多种果树和野花。这座花园让人联想到Jerez河的绿地景观（距离圣佩德罗小镇38km），并且体现了设计师希望保留沙漠环境中的稀缺绿地的想法。

在靠近酒店的公共区域，种植着大片的红色和金黄色奎奴亚藜。一条线形木栈道引导游人从这里走向场地的最深处，一直到场地的东北端。木栈道旁边种植着高大的玉米和向日葵，将这两种农作物或混合种植，或分开种植，大片玉米和金黄色的向日葵给人以静谧的感觉，让人真切地感受到安第斯沙漠的平静。

场地上种植着一棵牧豆树，树下摆放了几张木质桌椅，可供人们休息放松；采用传统技术修建的矮砖墙将整个空间围合起来，几扇篱笆门点缀其间，简单和谐的设计风格提升了圣佩德罗小镇的本土形象。

设计师对景观的深刻理解和巧妙运用使得该项目与众不同。该项目的很多局限之处都是自然环境造成的，但这并没有使设计师放弃对景观的改造或者将其转为他用；相反，这些局限成为了创新的机会，而场地中运用的景观设计技术已经被几代人实践过，并在其中融入了一整套当地特有的文化和风俗。这是有关土壤、水和自然循环的文化，有关土地、耕作和食物生产的文化，这片土地不仅是一处改良的景观，同时也是生成眼前一切景象的源泉。

1　酒店的一处水景
2　酒店客房
3　主入口
4　木栈道

Tierra Atacama is a hotel situated in San Pedro de Atacama, a Chilean town that was born as a natural oasis in the middle of the Atacama Desert. The site itself, which had been abandoned for thirty years, was restored as arable land and now appears as a purely productive landscape which responds not only to the hotel's own necessities but to those of the local community.

Situated in a place where water is a privilege and access to it becomes a matter of, first, intelligent planning and, second, this project shows how aesthetic objectives and functional ones can still live together.

Teresa Moller, the project's landscape architect, has found a poetic manner to go back to the roots of an agrarian landscape; historically developed as farming land and with the specific objective of producing the necessary food for the local community, the site is now bearing again its fruits. Moller decided to use the land according to its original purpose and, in addition, to create a place where colors, textures, perfumes and flavors combine finely; the fact that this creation takes place in the most arid desert on the planet turns it into a model of not only good management but also of humble acceptance of nature's given conditions.

San Pedro de Atacama is an oasis whose origin is directly related to the rains of the altiplanico winter, a climate effect typical of this part of the Andes which is generated by the Amazon's humid masses; the presence of the San Pedro and Vilama Rivers, originating in the Andes Mountains, becomes a decisive factor for the continued existence of this natural oasis.

The San Pedro River is the only source of water, other than underground water, for the homonymous town. When the river reaches the town it's channeled and distributed to the different working areas or "ayllus" into which the land is divided. These ayllus were the base of the socio-economic structure of these pre-Columbian human settlements. Today, the ayllus system is still active and San Pedro is differentiated into fifteen of them; however, native people have been moved to the periphery and the central areas have been occupied by hotels and accommodation for the present huge demand of tourists.

Based on a very rigid methodology and schedule, the water distribution system in San Pedro de Atacama becomes a key factor for land use. Every piece of land are irrigated by the manmade system of primary and secondary channels, every fifteen or eighteen days. These turns are the basis of the water management system. The actual irrigation technique consists of flooding the land until water infiltrates completely into the soil. The stipulated ration is two hours of watering per hectare, however, in those cases such as hotels, where the purpose of the land is not exclusively that of food production the final number depends on some other conditions; productive and social purposes and how old the site is are some of those particular conditions.

The site covers an area of 5.5 hectares and two secondary water channels run across the site. However, this technique has proved inadequate for certain crops, for it has deficiencies related to the impossibility of regulating the flood's physical limits and exact water flow-rate and volume. There are some parts of the cultivated land that receive less water than is required and some others where water runoff is quite high: water is either deficiently distributed or lost away. For this reason, the site had to implement other different irrigation systems, such as dripping, sprinklers and micro-jet, all of which respond to specific needs and complement the original method. The necessary water for these alternative techniques is obtained by means of a 60 meter-deep well built on-site.

The selection of the diverse crops that cover the hotel's land reflects that of the native crops that have been historically tested and developed in the Andean region. Corn, oat, barley and quinoa (the typical crop of the Incas) are the most representative ones and sunflowers, fruit trees and edible crops complete the extensive colorful list.

For the location of every one of them, two main parameters have been taken into account: first, a functional one which responds to the necessities of food production for the hotel; second, a formal one which responds to the site's general aesthetic. When closer to the hotel's installations the land's image is greener or more colorful, the more distant from the buildings and terraces the soil can get more exposed and the crops are less "tamed". On the west portion of the site, the fruit plantation and orchard offer a vibrant vision for the guests' rooms that face this area.

On the east side of the rooms' buildings, with the view of the mountains and volcanoes in the distance, some horizontally extended plantations of garden cosmos, a drought-tolerant plant with multiple tiny pink flowers, define a monochromatic carpet-like surface. Oat and barley are the species that alternate with garden cosmos, thus creating an attractive changing landscape throughout the year.

In the linear space defined between the east and west buildings, towards which all rooms have clear views, Moller laid out an intimate garden with edibles, fig trees and wild flowers. In a direct allusion to the Jerez Creek, a natural green formation developed along a brook 38km away from San Pedro, this place wants to preserve the idea of a sacred green within the context of a sandy desert.

When getting closer to the more public areas of the hotel, right next to the homogenous masses of grain crops, masses of quinoa display reddish and golden hues that mark the beginning of wilder and native species. From this area, a linear wooden path leads the visitor towards the last portion of the site; edging tall plantations of corn and sunflowers, which sometimes mix up and some others create uniform masses, this path follows the natural grading and reaches the

northeast tip of the land. After the imposing images of the cornfield and the garish yellow of the sunflowers' buttons, the crops seem to give place to a quieter more intimate spot where the peace of the Andean desert can be plainly experienced.

This peaceful spot is dominated by an old mesquite tree or algarroba underneath which the designer placed some wooden decks and seating surfaces where people are invited to rest and relax. The short adobe walls or "tapialeras", which were built with the traditional constructive techniques, close the space and alternate with a few fence-gates made of branches; the simple but consistent design of this area enhances the so emblematic vernacular image of the town of San Pedro.

What makes this project all the more remarkable is the use and understanding of a landscape whose many inherent constraints are actually imposed by nature. This landscape is not then left in sad abandonment or simply transformed into something else; on the contrary, those constraints are taken as new opportunities of creation. What's more, the techniques which are applied to get the most from the land are those tested by many generations, thus also incorporating a whole set of local traditions that constitute a specific culture.

It's the culture of the soil and the water and the cycles of nature; the culture of the land and the working of the land and the food production; the sowing and the harvest seasons and the season of the naked land, bare, raw, expectant of new crops and new fruits. The visitor, who comes from distant places and is trying to adapt mind and body to the height and to the desert, has also the task of learning that the land is not just that one of tamed landscape, cultivated soil and colorful fruits; the land is also the raw material that precedes all those images he certainly recognizes and understands.

1 篱笆门和矮砖墙
2 牧豆树和木质长椅
3 古老的牧豆树
4 收割过程中

景观与建筑的完美结合 —— 教育农场

A Remarkable Integration of Landscape and Architecture — Educational Farm

撰文：Jimena Martignoni　　图片提供：Barrado and Bertolino Office　　翻译：武秀伟

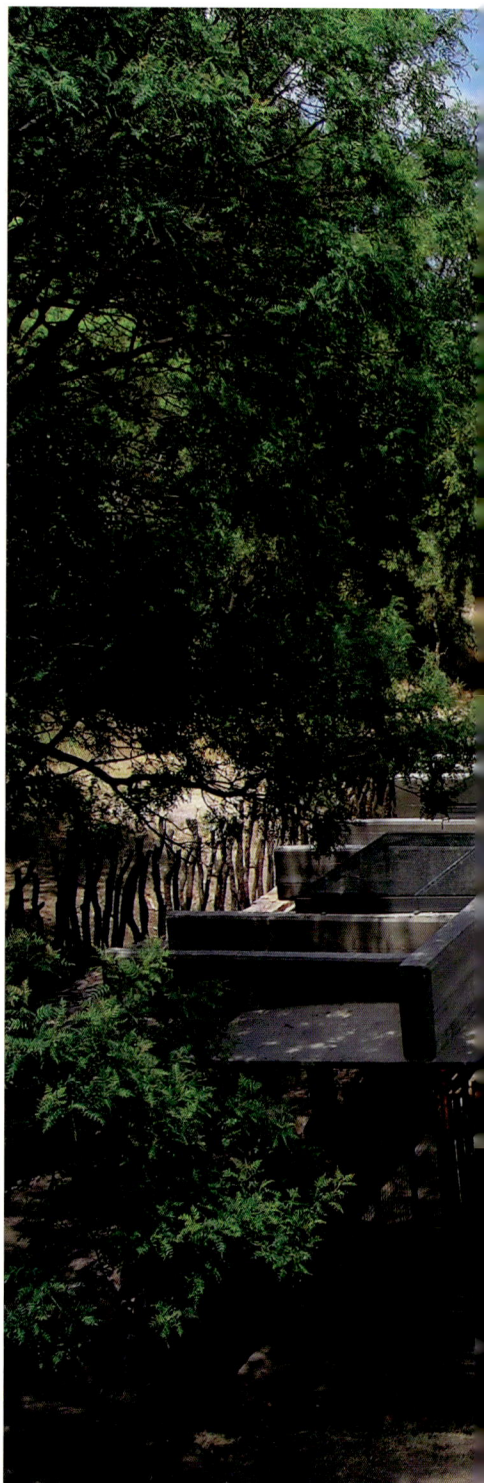

1 饲养棚架

该项目坐落在科尔多瓦市的干旱地区，是一个小型的、以食物加工为主题的农场。该项目是大型建筑群落中的一部分，是为了满足游客不断增多的需求而修建的。场地设计的初衷是要为游客们提供可以体验典型的农场活动，并了解食物加工过程的一处自然景观，加工后的食物都将提供给游客，满足了人们对食物及休闲活动的需求。

该项目占地 9650 m²，呈正方形。南面是陡峭的斜坡，上面种植着许多大树，东面是一座果园，北面是一条内部主街，西面是一座牧场。整个场地贴近自然且十分空旷，场地中分散种植着几棵古树，还有一处小坡度的斜坡缓缓延伸。

设计时将场地中现有的条件都考虑在内，以便保持场地的自然结构。场地入口临街而设，面向北面，周围巧妙地搭建了 50cm 高的线形石墙，顺着斜坡的自然走势形成一系列小巧的台阶，设计师利用这些台阶作为水景的载体。将水箱和水泵作为进水装置，同时，石墙起到了水道的作用。波光粼粼的水面和潺潺的流水声使人产生一种全新的愉悦感，并让人感受到一种特别的静谧，为这处干旱地区增添了视听魅力。儿童在这里玩耍时，流水声似乎可以让他们躁动的小身体安静下来。

设计师 Monica Bertolino 说："我们想把水资源引入到这个干旱的地方，人们很喜欢它，尽管它只是一处微小的景观……"该水道由当地专门从事石块建筑的工人修建而成，石块是从场地周围开采的，这种石块使设计显得更加本土化，并从另一个角度展示了铺设路面使用的石块的结构。该场地既展示了田园风情，又最大限度地体现了建筑设计风格。

根据项目要求修建了两处功能区：一处用来饲养动物，一处用来烤制面包和加工食物。设计师设计了两个矩形棚架，较大的棚架建在场地南侧的斜坡附近，与入口街道平行，用来饲养家禽以及农场中的其他动物；较小的棚架则建在果园旁边，用来放置烹饪器具。两个棚架采用相同的设计原则，既展现出自由的设计形式，又使构造一览无余。混凝土的主横梁和主立柱构成了两个独立的模块，与自然景观完美地融合。在视觉上，由于棚架四周没有墙，无论远处或近处的景致均衬托出独特的建筑结构；在结构上，因为屋顶并未被完全覆盖，树干和树枝都可以探出屋顶来沐浴阳光，一切元素构成了和谐的整体。

出于安全的考虑，饲养棚架必须进行围合。设计师在这里设置了木栅栏和铁丝栅栏构成典型的围栏，绵羊、山羊、小牛犊和大型禽类（如孔雀）都可以在

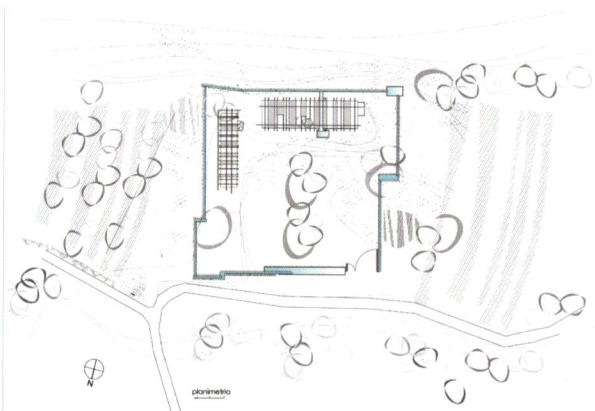

总平面图

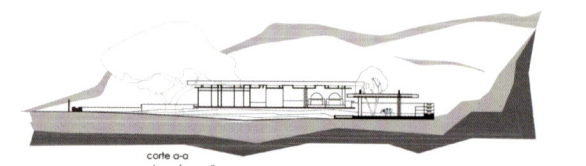

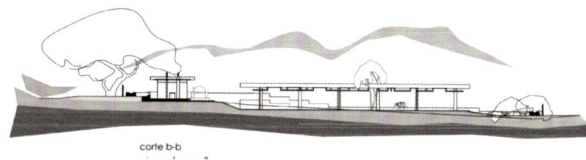

剖面图 1

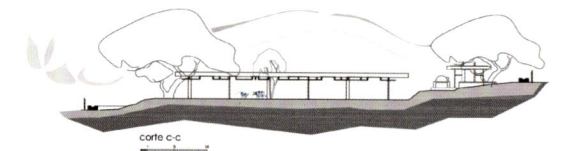

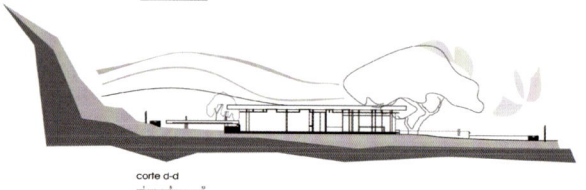

剖面图 2

1　棚架细部

2　原有的本土植物

半遮挡的棚架里饲养，饲养棚架的设计虽然普通，却不乏创意。在可供烹饪的棚架中，设计师们立起几块垂直玻璃板，以保持棚架内的卫生。最后一个棚架旁有两组泥土烤炉：一组放在架高的平台上，用来烤制面包；另一组放在棚架侧面的地上，开口朝上摆放，便于制作果酱。

场地上所有建筑结构以及泥土烤炉均为灰色——与绿色植物形成互补，两种颜色结合在一起有助于建筑物与植物相互融合。这种融合不单靠色彩来完成，建筑形体与大自然之间的对话也起了很大的作用——

屋顶的横梁与树枝一样参差不齐，延伸到建筑物外部。设计师有意通过这种设计方法在整个场地内形成光影斑驳的效果。树枝和混凝土立柱重叠或连接在一起，在地面上形成网格状的影子。场地的三面都用栅栏围合，将阳光作为艺术元素融入到设计理念中，光束透过栅栏，照射到旁边水道的石头上。制作栅栏的树枝和木条都是从场地上采集而来的，使得围合区域具有鲜明的本土特色。

与功能区相对的中心区域没有进行任何改造，保留了场地中原有的大片本土树木，如牧豆树、加州胡

椒木，在场地中形成了大片阴凉。烤面包棚架下的矩形方桌由设计师 Carlos Barrado 特别设计，每张桌子由一个水平面和两个垂直支架组成，桌子的外表呈淡红色，与原材料牧豆树的颜色相同。这些方桌采用了场地中废弃的木料打造而成，木片拼接的痕迹故意露在外面，鲜明地体现了加工过程中的循环再利用理念。

教育农场虽然形象朴素，但它以建筑结构为基础进行设计和建造却也不失田园风情，最终实现了天然原材料与精工细作的有机统一、形式与功能的统一，而最重要的是——景观与建筑完美地结合在一起。

The Educational Farm is a thematic and food production-oriented small project situated in the arid regions of the hills of Cordoba, in the geographical centre of Argentina. As part of a larger complex specially built to respond to the high demand from tourists, typical of the hills circuit of this province, the site was thought out with the specific objective of offering a natural place where visitors could experience the most emblematic farm activities and local food manufacturing. In addition, all food products of this farm would provide for the entire site and, in this manner, would meet a clear sustainable approach for a site with recreational objectives.

The project covers an area of 9,650 m² roughly shaped as an almost perfect square. Physically limited to the South by a steep slope planted with large trees, to the East by an existing orchard, to the North by one of the main internal streets, and to the West by a pasture area, the site is presented as a natural clearing only interrupted by old native trees and with a soft grading that descends from South to North and from East to West.

These existing conditions were incorporated into the project in order to maintain the natural structure of the site. The access was placed on the street, to the North, and the place's perimeter was subtly defined with a continuous linear 50 cm-high stone piece which, while following the naturally descending slopes, created a series of subtle steps.

The presence of these steps was part of the designers' decision to create a water-carrying element within the site. Fed by a water tank and a pump, the stone piece acts as a water channel or "acequia" that adds an attractive both visual and audile effect to this particularly arid place; the constant murmur of the falling water and the sun's reflections on the water surface all generate a new diversion for the senses as well as a special ambiance of tranquility. When kids visit the place, the sound of water seems to help to calm their usual vibrant behavior.

"We wanted to incorporate water as an element that the place lacks"—says Monica Bertolino—"Because people enjoy it even when it's a delicate presence…".

This acequia was entirely built by local workmen specializing in stone constructions and the stone used to construct it was extracted from the site's surroundings. This element helps to define a vernacular design and also alludes to the natural stone formations that frame the region's roads. However, the site offers a fine balance between this pastoral image and one that relates to the most representative elements of architecture and engineering.

In order to respond to the functional program of the project that requested a place for animals and another for bread making and food production, the designers proposed two different rectangular pavilions positioned perpendicular to each other. The larger one, which is meant to accommodate livestock and other farm animals, is placed adjacent to the natural slope on the South side of the site and parallel to the

accessing street; the smaller one, which provides the cooking infrastructure, edges the existing orchard.

Both pavilions respond to the same design principles of a free plan and structural systems exposed to view. Concrete main beams and columns make up two independent modules which have in common the particular attribute of letting nature grow through them. Either visually, because the absence of walls lets close and distant vistas appear as part of the architectural composition, or physically, because the absence of solid roofs lets branches and entire trees reach out for light and go through them, all pretending to be part of a single composition.

In the pavilion for animals, where it was essential to close the space for evident safety reasons, the designers created some wooden and wire fences that want to recreate the typical concept of a livestock corral; they're innovative yet plain designs. Ships, goats, baby cows, and some large bird species such as peacocks share a single semi-roofed structure which doesn't compete with the natural setting. In the bread making and cooking pavilion the designers placed just some vertical glass surfaces in order to maintain clean and hygienic conditions.

Next to this last pavilion, there are two sets of mud ovens: one set for bread baking and the other for jam making. The first two are placed just in front of the cooking tables' area on an elevated platform to allow easier access for bread handling; the other two mud ovens are placed on the ground, at one of the pavilion's sides. The mouths of these last ones are on the top to allow an easier stirring process.

Altogether, the grey hue of the concrete—of every one of the structural elements and also the mud ovens—finely contrasts with the green tones of the foliage and generates a color combination which helps to define the expected integration between buildings and plants. This integration, however, is achieved not only by means of color but also through the respectful dialogue established between the shapes of architecture and those of nature; the roof beams, whose length varies while overhanging out of the structure, emulate tree branches that grow and extend dissimilarly. The usual need of precision or exactness of structural engineering pieces gives way to nature's unpredictable rules.

This play of elements and dissimilar pieces was also intentionally laid out to create shade and sunlight reflections throughout the area; branches and concrete beams seem to overlap and interlock to shape light and shade networks onto the ground.

The design of the fence that was built along three of the sides of the site's perimeter (the North side which coincides with the access presents a more formal layout) also complies with that same concept of using sunlight as an aesthetic element that can be easily incorporated into the plan; in this case, the sunbeams that go through the sticks—with which the fence was built—are projected onto the stone surface of the acequia that develops right next to it. Built with bough

pieces and sticks found on the site, this fence offers another reference to the vernacular character of the project and the area in which it's enclosed.

The central space, towards which open up the two pavilions, was left as an untouched space where a few existing clusters of native trees such as prosopis or mesquite tree and schinus molle or peppertree species provide large shady spots.

The furniture of the bread making-pavilion, composed of fine rectangular tables, was specially designed by Carlos Barrado and built with wood scraps from the construction site. These tables were built with the specific objective of providing a surface for bread kneading and the design responds to a clear recycling process-based concept.

The wood that was utilized for the construction of the tables was supplied by discarded cuts of parquetry or floorboard in the site; the pieces and the manner in which they were engaged are intentionally exposed to view, thus emphasizing the idea of reutilization of existing materials. Made up of one horizontal plane and two vertical bearing ones, these tables display an eye-catching reddish tone typical of the prosopis wood with which they were built.

The simple yet accurate image of this didactic farm and the structurally-based yet bucolic design concept with which it was outlined end up in a remarkable integration of use of raw materials and finishing techniques, form and function and, above all, landscape and architecture.

朗根洛伊斯葡萄园露台

Vineyard Gazebo of Langenlois

撰文 / 图片提供：Aubck+Krsz　　翻译：李沐菲

1

该项目是为庆祝 2006 年下奥地利州的园艺节而建的，周围的通道区于 2009 年完工。奥地利雕塑家黑默·左伯尼格设计的雕像与设计师的努力共同打造了这个具有浓厚艺术气息的葡萄园。

在过去的十年间，现代葡萄园建筑成为了建筑产业中很有吸引力的一个主题。然而，极少有项目是从景观设计这一角度出发来设计建造的。由建筑师史蒂文·霍尔设计的著名的"Loisium"酒庄就坐落在不远处——这座乡间小镇的边上。为了受邀前来参加独一无二的世界葡萄酒之旅的众多客人，当地建造了许多酒窖，形成一个可追溯到九百多年前的迷宫，而"Loisium"酒庄与其他众多酒庄一样，为数个葡萄酒生产者提供服务，而露台则属于奥地利著名葡萄种植园之一——布德梅尔酒园。它既是欣赏葡萄园美妙景观

的一个绝佳地点，同时也为特别的客人和特别的宴会提供了一个相应的户外空间。

客户的想法是要设计一个轻型的建筑结构，并能够同这里极具象征意义的环境建立起一种特殊的联系：一个从长远看近似于无形的实体建筑。

该项目包含了三个元素：即与葡萄园中小路相连的长长的通路、露台本身，以及与黑默·左伯尼格的雕塑之间的联系。

游客们沿着一条狭长的通路进入这里。通路一侧是新建在使用传统技艺建造的葡萄园中的干式墙——用以加固陡峭的斜坡；另一侧则是一排葡萄藤，而从一开始就成为视觉焦点的雕塑便是这条小路的尽头。

虽然露台本身与雕塑距离很近，但只有在到达平台高处的时候才能够看见雕塑。它坐落在两个葡萄园

之间的斜坡上，具有建筑和景观的双重功用。站在葡萄园上部的木质平台，即观景台，能够俯瞰这座山谷的全景，而众多的小葡萄园便是在这里生活的根基；而露台下层则采用了景观设计中传统的框式景色，给游客带来更亲近、安全的感觉。在这里，酒庄向尊贵的客人展示他们的产品，山上还修建了酒窖，以保存他们的葡萄酒。露台顶部覆盖着蜿蜒的葡萄藤，使其完全融入到周围的田园风光。

摒弃建筑物的外观、却又不失其与游客之间交流互动之功用的设计理念最终在这里得以实现。

与此同时，除了史蒂文·霍尔设计的"Loisium"之外，这一小而隐蔽的标志性建筑也为建筑师和景观设计师在设计上提供了一些建议，难于发现却又乐在其中。

The Gazebo in the vineyards of Langenlois was built as a contribution to a garden Festival in Lower Austria in 2006 and completed with its surrounding access area in 2009. It is part of an artistic intervention together with the Austrian sculptor Heimo Zobernig.

Contemporary architecture for wineries became an interesting topic in the architectural production of the last decade. However, there are only very few projects which were created from a landscape perspective. The well known "Loisium" by architect Steven Holl stands in sight distance on the fringe of this small rural town.Its wine cellar world is a labyrinth dating back 900 years which goes through the old wine cellars of the town of Langenlois and invites the visitors to participate in an exceptional tour into the world of wine. While the "Loisium" serves amongst others as a vinotheque for several winegrowers, the gazebo belongs to the winery Bründlmayer, one of Austrias most famous vine cultivators. It serves as an intimate location with a marvellous vista within this landscape of vine, as an outdoor space for special guests, and specific feasts.

The goal of the client was to realize a light construction that establishes a particular relation to this emblematic ambiance: a piece of architecture which does not appear as such, which becomes, in the long run quasi invisible.

The project enfolds three elements: the long access from the small road amidst the vineyards, the gazebo itself and finally its relationship to the sculpture of Heimo Zobernig.

Visitors approach to the site along a narrow passage between a new drywall, built in the traditional technique of

1 露台细部
2 建成后的露台
3 设有酒窖的露台

the vineyards for securing the steep slope and a new row of grapes; the sculpture marks the end of this path being in the focus right from beginning.

The Gazebo itself stands close to the sculpture, but in a distance of respect, to be seen only when arriving to the plateau. Situated on a slope connecting two vineyards, it forms a symbiosis between Architecture and Landscape. The upper wooden deck, a kind of belvedere, offers a panoramic view over the valley, where life is based on small viniculture units. The lower level is characterized by a framed vista, a traditional motive in the history of landscape design. Here the visitors feel more intimate and protected: This area serves for the winegrower to present his products to special clients, for whom the vine bottles are kept in the cellar constructed in the hill. The gazebo, covered by climbing grapes will gradually merge with its surrounding agrarian environment.

The idea to create an object, which shall disappear in its architectural appearance but keep its perceptive and communicative value for the visitor is on its way to be completed.

Beside the "Loisium" of Steven Holl this little hidden landmark meanwhile became an insider tip for architects and landscape architects, difficult to find, pleasant to enjoy.

1 从露台向城镇眺望
2 露台上层与雕塑
3 露台底层
4、5 从葡萄园内观赏露台
6 入口通道

新兴的城市农艺 —— 美食庄园

Emerging Urban Agriculture — Edible Garden

撰文：ARTECHO　　　图片提供：Steve Gunther　　　翻译：谷晓瑞

在美国，农产品从产地运到餐桌上平均要走约2000千米至2600千米的路程。而在克里斯·萨帕诺丽这个位于加利福尼亚州圣塔·莫妮卡市的小镇则仅需约4.88米。

在圣塔·莫妮卡，私人庭院中的草坪对水的需求量很大，ARTECHO建筑与景观设计事务所的景观设计师帕梅拉·帕尔玛带领她的团队采取节水和被动冷却措施，将草坪变成了美食花园。当一个城市中的空间

和水都成为了重要的商品且气候又适宜农作物四季生长的时候，人们将草坪转为种植农作物就不难理解了。虽然在私人庭院中种植农作物已不是什么新鲜事，但将整个庭院种上农作物的做法还是一个创新。随着市场对本地农作物需求的不断增加，也许将会有更多的业主在庭院里栽种绿叶作物。

以往的草坡不见了，在重新规划过的庭院四周垒起了石墙，石墙旁边搭起一个花藤来为房子遮阴。一

条由碎石铺就的石阶通向庭院，点缀以一个混凝土砌成的喷泉。这里还设计了一个石头露台，人们可以在这里休息或在花藤下乘凉。美食花园四周种植的是耐旱的非农作物植物，它们虽不能食用，但却形成了一道艳丽的风景线：鲜艳的紫色墨西哥鼠尾草与耀眼的橙色澳大利亚针叶植物争鲜斗艳。穿过洋蓟区、茄子区和果树区，就到了平坦的庭院入口；苹果树和无花果树沿着房屋的外墙生长，葡萄在藤架上蔓延，人们

站在颜色鲜亮的前门就能闻到柠檬的香甜。

在房屋前的院子里有五个高身花槽，这是花园中最吸引人眼球的地方。水果和蔬菜在特制的花槽中生长，花槽里不仅设有一个长凳，还是存放有机肥料以及工具设备的地方；人们可以在长凳上休息、品尝收获的时令果蔬。邻居们可以在这里小聚、除草，抑或是吃草莓、罗勒和番茄。这些特制的高身花槽和整个美食庄园都采用滴灌系统，以确保水流能浇灌到植物根部而不会被浪费。在加利福尼亚南部，水分蒸发很快，因此相较于过去的喷灌系统这种方法就非常节水。

设计师充分利用了转角地段，在庄园的周围种植了很多食用果蔬。设计师沿着栅栏种了五棵果树，果树之间生长着西瓜和当地的草本植被。路人在街上就能看到像足球一样大的西瓜，这在加利福尼亚南部的私人庭院里是很罕见的。葡萄藤、无花果树、苹果树和樱桃树，它们既可以结果实，又能为房屋遮阴，减少了空调的使用，真是一举两得。入口处设有石阶和石廊，在石廊上爬满了藤蔓，最适合邻居们相互走动、分享收成。设计师的设计集可持续性、健康性、节俭性、疗养性和娱乐性于一体，该项目也成为了未来城市农艺的典范。

In the United States, the average distance produce typically traveling from field to kitchen is between 2000 and 2600km. For Chris Spagnoli of Santa Monica, California, the distance is only 16 feet.

Landscape Architect, Pamela Palmer and her team at ARTECHO Architecture + Landscape Architecture, transformed this Santa Monica resident's water thirsty front yard lawn into edible garden using water conservation and passive-cooling practices. In a city where space and water are precious commodities, and the climate is ideal for year-round produce, exchanging turf for food makes simple sense. While growing food in residential gardens is nothing new, this concept of transforming an entire front yard to accommodate edibles may be considered progressive. With the increasing demand for local produce and the rising trend of farmer's markets, more homeowners may be switching out their front lawns for leafy green edibles.

The once sloping lawn has been removed and a stone wall built to retain the reconfigured yard. A pergola has been added to provide shade for the house and stone entry steps built which lead to a gravel entry court punctuated by a cast concrete fountain. A stone terrace has been built where one can relax or visit under the shade of the new pergola. Bright swathes of non-edible, yet low water plants line the edges of the property and contain the edible garden. The bright purple of the Mexican sage (Salvia leucantha) contrasts with the stunning orange Australian pincushion plants (Leucospermum). One walks between artichoke, eggplant and squash plants to arrive at the level entry court. Apples and figs are espaliered on the walls of the home, grapes climb an arbor and the sweet scent of

lemon greets one at the brightly painted front door.

The focal point and neighborhood gathering spot of the garden is the area in the front yard featuring five raised planters. Fruits and vegetables flourish in the custom planters designed with a built-in bench and storage space for organic amendments, tools and equipment. The bench provides the perfect place to sit and harvest whatever produce may be in season. Neighbors meet and weed or nibble on strawberries, basil and tomatoes. The custom raised planters and the entire edible landscape are watered with a drip irrigation system constructed to ensure water gets to plant roots and is not wasted elsewhere. The drip irrigation system constitutes a substantial amount of water savings over the previous spray irrigation system, where water quickly evaporated in the Southern California sun.

Taking advantage of the corner lot condition, the landscape architect was able to plant more edibles along the perimeter. Five fruit trees are planted against the fence with melons and native grasses planted between them. Passersby on the street have eyed soccer ball sized watermelons; a rare view in front yards in Southern California. Grapevine, fig, apple, and cherry do double-duty providing food as well as shading the house, and minimizing the need for air conditioning and saving energy. The entry area with its stone steps and the stone porch with pergola, is the perfect place for visiting with neighbors and sharing the harvest. The ARTECHO designed garden is a great example of the potential of urban agriculture. It's sustainable, healthy, money saving, therapeutic and fun.

伦敦市新的创意平台 —— 温室

A New Creative Powerhouse for London — Hothouse

撰文：Pedro F. Marcelino　　　图片提供：Ash Sakula 建筑师事务所 Cany Ash　Fredrik Rissom　Ioana Marinescu　Nick Guttridge　　　翻译：武秀伟

从 20 世纪 80 年代开始，自由形式艺术机构就纷纷在伦敦许多小型场地上开展"艺术存于心中"的城市重建工程，以其独特的使命感为艺术及创新产业提供服务。近年来，该机构在环境改造方面扮演着越来越重要的角色，在伦敦的各个角落实施许多宏伟的重建计划，包括未来奥运场地的重建工程。新的创意空间是该机构更好地开展创造性工作的必

备条件。该项目是该机构的新家，地处伦敦东部的 Hackney 区，能够满足这种需要，因此，大批从事公共空间改造的艺术家、设计师及建筑师正聚集到该创意空间中。

该项目将为建筑师们带来很多灵感。Hackney 区内的很多建筑都需要重建。自由形式艺术机构搬迁到该区域附近，虽然不能解决该区域面临的许多严

峻的社会、经济和城市规划问题，却可以使更多有创意的建筑师们零距离地了解该区域的现状。该项目位于运动场和繁忙的铁路高架桥之间（连接利物浦街道车站和 Stansted 机场），这也体现了该区域在城市中所处的重要地位。

正如人们常说的，项目可以为自己寻找到合适的建筑师，这一小片凌乱的星形场地恰好验证了这

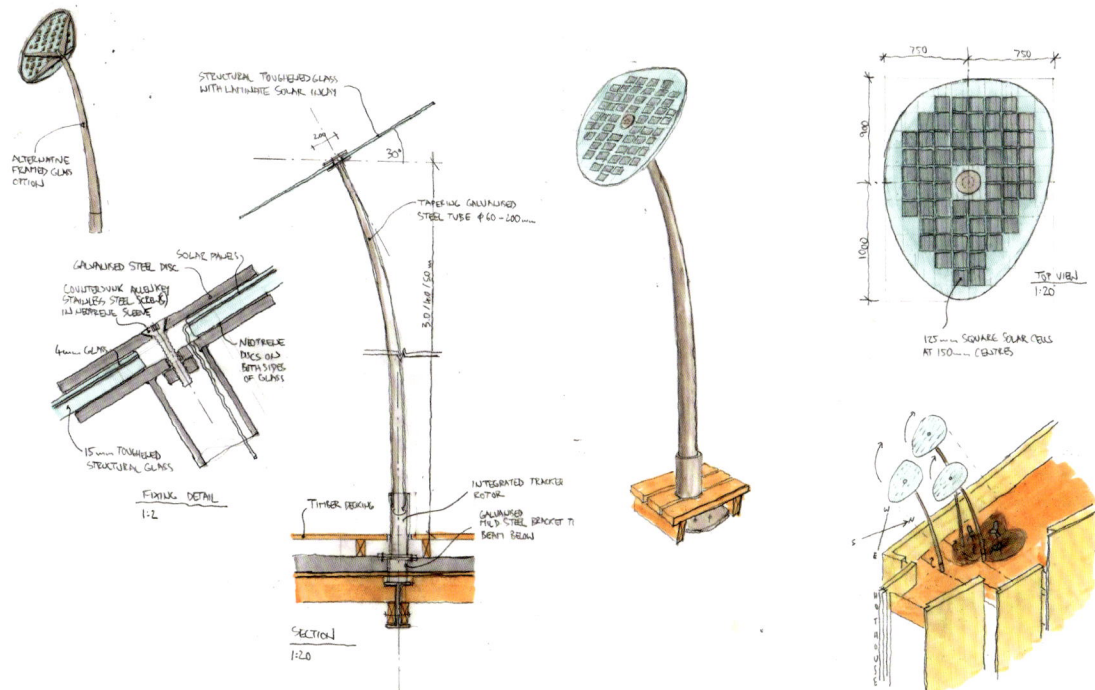

太阳能装置嵌入详解

1　回飞镖形状的建筑
2　当地的孩子们在参与栽种植物

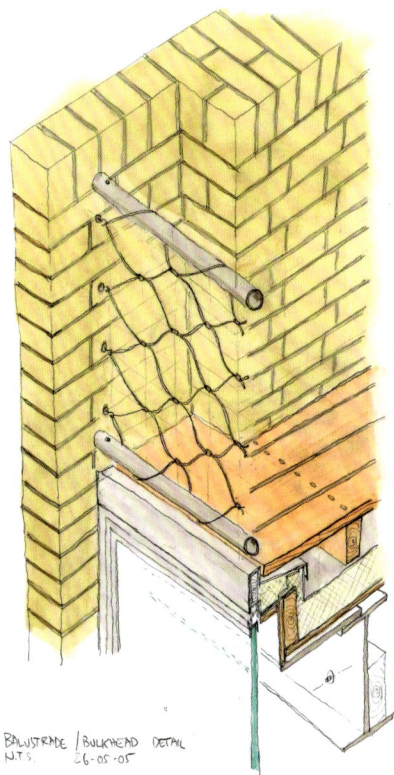

扶手细部

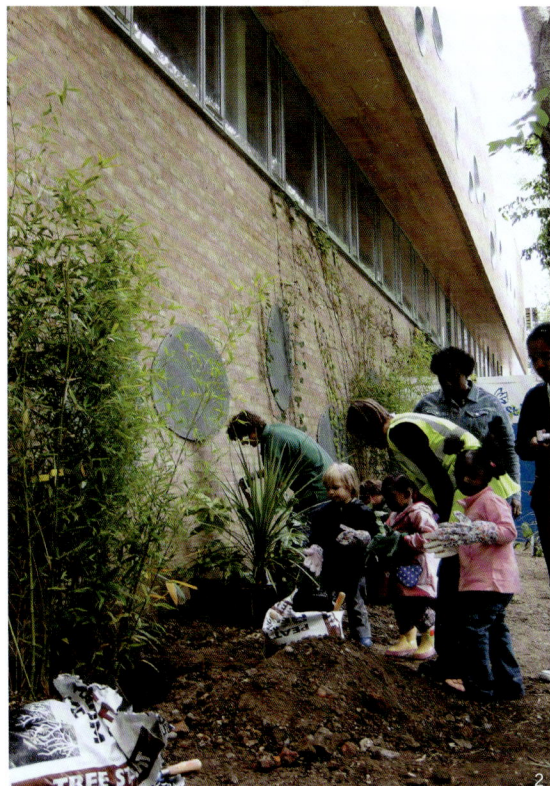

种说法。Ash Sakula 建筑师事务所接受挑战，并担负起 Hackney 区的改造与重建工作，它是伦敦的一家专门从事建筑改造与重建、艺术设计和创新设计的工作室，了解整个项目的要求后，设计师在原有的基础上对其进行了改扩建设计。原本让人们退避三舍的破败区域如今成为了最热门的建筑项目所在地，该项目还获得了 2003 年度两项备受人们瞩目的伦敦规划奖。

富有创意的回飞镖形状的建筑与原有的背景交融在一起，使其重新焕发出活力。建筑外围砖墙的弯曲形状也向人们展示了其传统的风格，建筑物立面散布着许多阿米巴虫形的窗户，它们是根据客户要求特别定制的。这些窗户也许会成为该建筑最吸引人的特征之一——从室外看，它们使得面向操场的外墙充满趣味，并且与操场的景观相协调；从室内看，窗户似乎将花园的风景定格，使风景几乎走进了办公室。该建筑中的一些区域拥有多方位的视角，人们可以透过四扇不同的窗户看到同一棵树的不同侧面，好像树木就生长在屋子里一样。光影效果形成了另外一种有趣的景观，但这并不妨碍艺术家和建筑师们坐在办公桌前为伦敦市构思新的改造规划方案。

呈几何形状的中央木质楼梯是该建筑中另一处引人注目的景观，或许也是整体设计中最成功的一部分。楼梯含蓄地将室内富有创造性的工作与室外瞬息万变的城市联系起来，楼梯的尽头成为建筑与周围环境互动最关键的部分。设有照明设施的屋顶露台为艺术家及自由形式艺术机构的工作人员提供了交流、切磋的场所；露台也能增进建筑与环境的互动——一辆急速行驶的列车从几米之外驶过，把城市的景观带到屋顶，而屋顶某个短暂的艺术瞬间又被带回到列车上。更确切地说，是带到了伦敦市民的心中。

Ash Sakula 建筑师事务所的设计总是把环境因素放在首位——若要使建筑具有持久的生命力，不能只依赖于选择一些绿色材料，而是应该意识到，建筑在本质上是反生态的，应将环境因素纳入到整体设计中去，从原材料的选用到公共系统整体功能的发挥都应如此（例如利用光电池来发电）。初看该场地比较朴实，但该项目不仅是自由形式艺术机构的新家，也是伦敦一项富有创意的新资产，是客户与设计师圆满合作的成果。

1　创意动力体现在活动中
2　光影效果将室外景观引入室内
3　花园
4　阿米巴虫形的窗户

3

4

Since the 1980s the Free Form Arts Trust had been operating its 'art at the heart' of urban regeneration projects from a myriad of small sites spread across London in its unique mission of providing services for the arts and the creative industries. Catering specifically to interventions in the built environment, this role has been all the more important in recent years, with a number of ambitious regeneration plans scattered throughout the city, including the massive re-conversions taking place in the future Olympic sites. It was a logical development that the Trust would eventually need to settle down in its own purpose-built space, conducive to the creative work involved in any of its projects. Hothouse, the Trust's new home in Hackney, East London, provided an answer to that need, and is quickly becoming a creative cluster for arts, design and architecture professionals working on public space

interventions.

The site could not be more suggestive, as far as work inspiration goes. Hackney has long been an area of London in dire need for re-qualification projects of different sorts. Bringing the Free Form Arts Trust to the neighbourhood certainly did not fix the severe social, economic and urbanistic problems the area faces but it did put the creative brains inside the Hothouse in flagrant contact with the reality of this community. The location of the terrain nestled right between a playground and a busy railway viaduct connecting Liverpool Street station to Stansted Airport is a further reminder of its eminently urban condition.

As the saying goes it is the project that looks for the architect, and this little star-shaped site found itself a small practice that could. Ash Sakula Architects, a London-based studio specializing among other things in interventions in

the fields of regeneration, the arts and creative industries, and the firm indeed took the challenge, understanding the brief and pushing it beyond its borders. All of a sudden, this sorry little piece of Hackney ground shunned by major architects became the hottest project of the moment, receiving two of the coveted new London Planning Awards in 2003.

The imaginative boomerang-shaped building dialogues with the pre-existing surroundings but also offers them something new and exciting. The building's brick envelope remains as a reference to the area's traditional architecture, although its curvy shapes are all but orthodox. The façade, however, is peppered with amoeba-shaped custom-made windows. These windows are perhaps one of the most striking features in the building: from the outside, softening the wall that faces the playground, and indeed looking so fun that they could be a part of it; from the inside, they frame the garden in such a way that it is nearly brought into the office space. In parts of the building, sections of a single large tree are visible through four different windows, offering a multitude of perspectives and nearly suggesting that the tree is in fact indoors. The light and shadow effects provide a further point of interest that certainly is not detrimental to the artists and architects sitting at their hotdesks creating vibrant new interventions for the city of London.

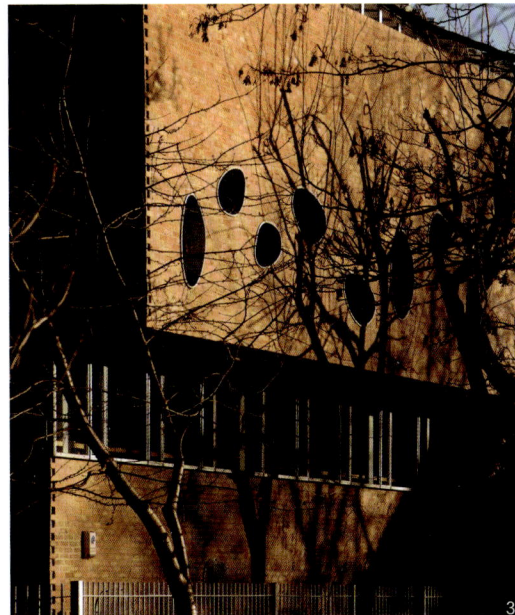

The building's woody central staircase is a masterpiece of geometric playfulness, perhaps one of the most well achieved elements of the overall design, and metaphorically connecting the creative work inside to the transitory city outside. In terms of the interaction between the building and its environment, it is to where the stairs lead that is the most important. A cleverly lightened rooftop terrace stretches atop the building as an inviting new performance site for the artists working with Free Form Arts Trust: it is outdoors, exposed and living. So living, in fact, that in rush hour a speeding train passes a few meters away at eye level, bringing the city—onto the site. Or indeed, bringing a short-lived art intervention from the rooftop onto every passing train, and indeed to the people of London.

As with many Ash Sakula projects, the environmental aspect was not ignored, rather taken onboard. The firm's stance that the best way to achieve sustainability is not simply to choose green materials but to recognize that construction is in essence a counter-ecology meant that environmental concerns were mainstreamed into the design, from the materials to the global functioning of utility systems (e.g. photovoltaic cells provide energy to the site). Humble as the site may have seemed at first, the Hothouse is not only Free Form's new home, but also a new creative asset to London and a great testimony to a fruitful collaboration between client and designer.

活历史 —— 多里斯农场总体规划

Living History — Dorris Ranch Master Plan

撰文：Allison Collins

图片提供：Courtesy of Sally McIntyre; Courtesy of MIG,Inc.

　　　　　Courtesy of Brett Cole; Courtesy of Laurie Matthews and Christina Frank

　　　　　Courtesy of Williamalane Park and Recreation District

翻译：张晶

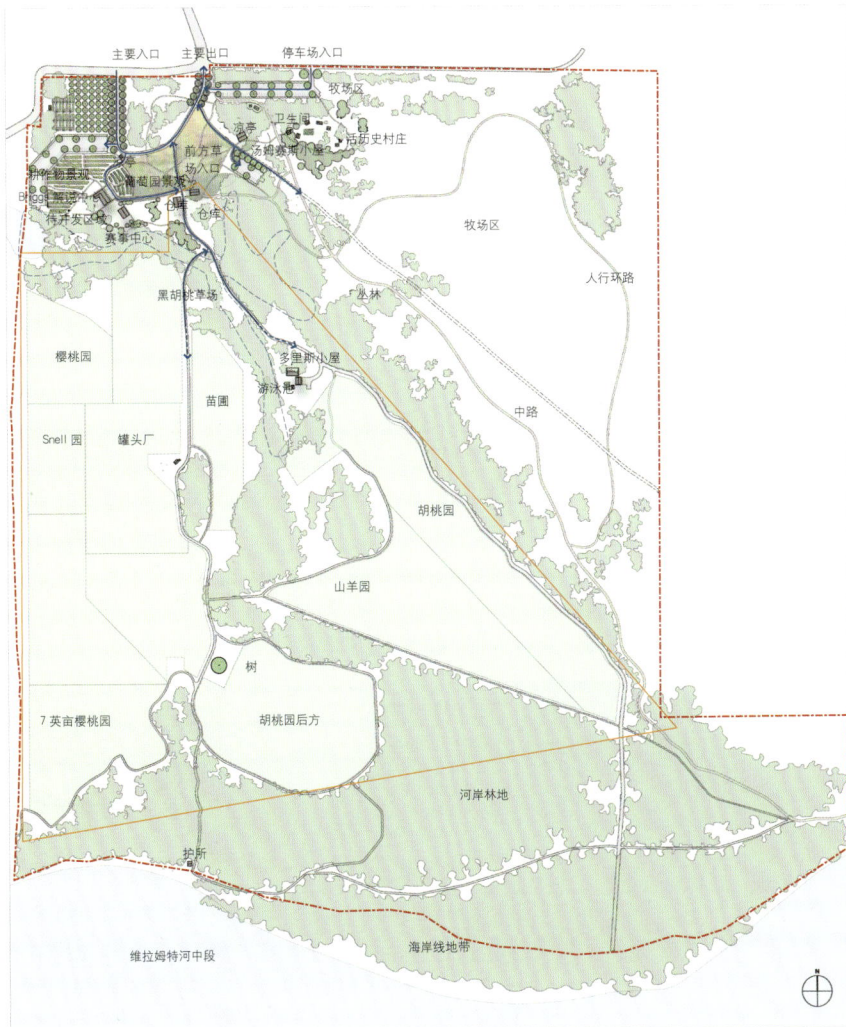

对于一个农场来说，如果能够入选美国的《国家史迹名录》（National Register of Historic Places）的确是很大的殊荣，但同时也会带来许多限制，提到改变，人们都会摇头说不。多年以来，多里斯农场一直受到种种限制，几乎成为保守思想的牺牲品。然而，现在流行复古了，设计组发现该项目是一项可行性的、充满无限可能的、内容丰富的工程——可以将美国最古老的商业性榛子园保留下来，对其进行主动式管理；可以提升游客中心的服务质量，增加人们喜爱的园区小径路线；还可以开辟出更多的空间和教育场地用于出租，供人们举行各种各样的活动；当然，也可以保护和改善稀有

的湿地、橡树林和草原动物栖息地。

该项目的总体规划给设计带来了许多挑战，同时也带来了更多的机遇。首要的挑战就是需要确定这个特别的公园的游客接待量。在保持20世纪初农场原貌的条件下，展示优美生态的系统及其重要性的同时，增加哪些服务能够吸引更多的游客呢？农场的环境条件以及原始文化、自然特色的保留的确会给该项目的开展带来各种阻碍，但也带来了机遇，如在同一场地内融合建筑设施、历史和自然等多种元素。可以说，在俄勒冈州乃至美国，极少有如此多元化的场地。

该项目的生态多样性和文化重要性引起了许多

令人关注的开发与管理问题，因此针对农场的实际特点及其要发挥的体验价值而制定长期规划是极有必要的。该项目最新的总体规划就是要将俄勒冈的早期历史、农业景观、乡土建筑、社区建设以及稀有的自然资源完好地保留下来，并对其进行适度的扩充和完善。

该项目在规划和设计的过程中，设计组高度重视通过网上的民意调查、股东见面会、现场办公等方式收集民众意见，并在项目指导委员会的指导下明确了设计方向。多年以前，曾经因为几个有争议的问题而阻碍了多里斯农场规划的实施。但该设计规划方案最终获得了人们的一致认同，在未迫使任

何人做出让步的情况下解决了所有实质性问题。

　　设计组清晰地认识到这些问题都不是独立存在的，他们知道保持该项目文化和自然特色的关键就是要平衡历史与现代的管理。为了保护和完善多里斯农场的特有风貌，并将其留给子孙后代，该项目的规划方案设置了以下四个重要步骤：

1. 保障榛子园的运营

　　该项目的核心就是其农业历史。从体现该农场的历史完整性上来说，持续而积极的农业生产的意义要大于其他任何一项单一行为。因此，积极地种植榛子树和收获榛子对保持该农场的历史特色是至关重要的。然而，设计组面临两个较大的挑战：一是，在农场东部榛树枯萎病蔓延时期，因事先无法通知到每个农场，所以导致大部分园区必须关闭，园内的一条主干道也将被阻断；二是，园内西北角新种植的大片榛子树便成为了传播病害的途径。考虑到这些问题，设计组建议通过移走稀疏分布的榛子树来控制病害的蔓延，并开辟了一条新的路线，可供游客在疫情发生时前往其他未感染区域游玩。

2. 减少交通冲突

　　农场里经常会发生人与车辆的交通冲突，主要地段是人车共用的入口前方道路以及散乱分布的非正式停车场。为此，设计组特别设计了环形入口，铺设了直通停车场的小路，该停车场设在游客活动中心区之外，避开了人们的视线。另一处停车场则

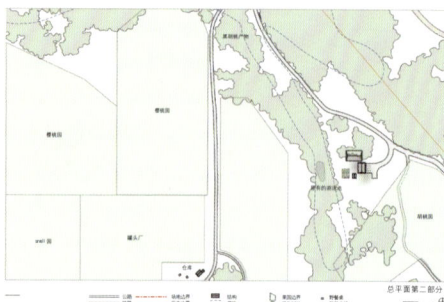

隐藏在农场北缘，邻近地区公路口，既满足了大型活动的泊车需求，方便游客就近前往附近的公路，同时也减少了这些游客对历史中心区的影响。设计组为了更加长远的打算，在进行场地分析时还提出了备选方案，即将入口处的小路弃置不用，重新启用农场西边原有的一条道路。

3. 增加教育培训和各种活动

　　社区居民和公园管理人员期待能有更多的教育实习及其他一些大型活动在这里举办。但是，他们更希望在吸引游客的同时能够将由此对文化和自然资源造成的影响最小化。举行婚礼及特定活动的场所已经确定并预留了出来，而且还增加了一些设施来方便使用和保护资源。专门用来举办教育性活动的仓房和"汤姆塞思之家"（Tomseth House）也可以出租举行其他室内活动。除此之外，设计师还增加了一处新的教育及活动场所——处于停车场入口与老果园和小路之间的中心点，从那里可以眺望正前方入口处的大草场。

4. 保护珍稀的天然栖息地

　　橡树、大草原和河岸栖息地一度是非常重要的文化、自然生态系统。在威拉梅特谷（Willamette Valley）发生的几起大规模火灾中，美洲印第安人将橡树和大草原保留了下来。可是，如今橡树和草原的面积却越来越小，动物的栖息地也几乎消失。保护只是保留这些重要自然系统的第一步，而目标性管理技巧才是保障其持续健康发展和复兴的重要手段。设计组不仅对大面积的橡树林、草原、林地和河岸地区采取了保护措施，而且还同管理人员共同制定了循环保养法，以保障这些珍稀栖息地的持续健康和蓬勃发展。

While it's an honor for a site to be listed in the National Register of Historic Places, that designation often comes with a perception of limitations. NO becomes an expected refrain. Through years of focusing on limitations Dorris Ranch was close to falling victim to the NO mentality. Instead the tide turned. The planning team saw this as a project of YES, a project of possibilities, and a project of inclusion. Yes, we can preserve the oldest commercial filbert orchard in the United States and keep it in active management. Yes, we can increase the level of visitor services and expand the much loved trail network. Yes, we can create more versatile event rental spaces and educational venues. And yes, we can protect and enhance the rare wetland, oak woodland and prairie habitat.

The Dorris Ranch Master Plan Update provided the planning team with many challenges, but also many opportunities. Challenge number one would involve determining the carrying capacity for this unique park. What types of visitor services could be added to serve expanding interest from the community while still retaining the character of an early 20th century farm and the functionality of a delicate and important ecosystem? Site conditions and sensitive cultural and natural features provided the constraints, but also presented an opportunity to integrate facilities, history and nature in one place. There are few sites in Oregon, or even the nation, that have this combination of elements.

Dorris Ranch is rich in ecological diversity, and its cultural significance raises interesting development and management issues that require a long-term plan, sensitive to the authentic character of the site, as well as its experiential values. The purpose of the Dorris Ranch Master Plan Update was to preserve and appropriately enhance the distinctive and vital blend of early Oregon history, agricultural landscape, vernacular architecture, community programming and rare natural resources that exist on the site.

Throughout the planning and design process the planning team prioritized this information based on the input they received from the public through an online survey, stakeholder interviews, and an on-site open house; and honed the direction with the guidance of the project steering committee. Several conflicts had held up planning for Dorris Ranch in previous years. This planning team was able to achieve consensus where others had failed—crafting a plan that addresses all pertinent issues without forcing anyone to compromise their point of view.

The planning team understood that these components are not mutually exclusive of each other. They determined that the balance of historic and contemporary management practices would be critical to maintaining Dorris Ranch's cultural and natural characters. This plan took four primary steps to ensure that the very essence and character of Dorris Ranch would be protected and enhanced for current and future generations:

1. Maintain a Working Filbert Orchard

At the heart of Dorris Ranch is its agricultural history. The fact that active agricultural production has continued on this site has contributed greatly, perhaps more than any other single action, to the site's historic integrity. Therefore, preserving active filbert farming and harvesting

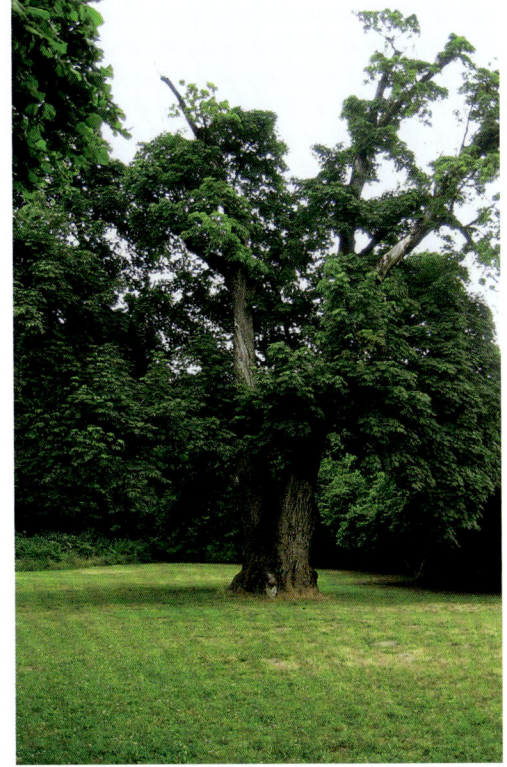

is vital to the preservation of the site's historic character. However, two large challenges emerged. One, that a large portion of the park needs to be closed without a great deal of advance notice when spraying for Eastern Filbert Blight occurs effectively cutting off a major trail network that is one of the site's most heavily used features. Two that a grove of non-historic filbert trees existed in the site's northwest corner and could serve as a conduit for disease to reach the historic orchards. With this in mind, the planning team recommended the removal of the stray orchard so spraying could be more contained, and developed an alternate trail network that would provide visitors access to portions of the property not closed during spraying events.

2. Reduce Circulation Conflicts

Vehicular and pedestrian circulation conflicts were spread throughout the site, primarily by an easement road, through a joint entrance/exit, and a sprawling and undefined parking lot. The planning team eliminated these conflicts by designing a loop entrance marked by an allee that leads directly into a parking lot that is located outside the core of the visitor area and tucked out of view. Another parking lot was tucked along the site's northern edge and near a regional trail head. This lot will provide overflow parking for large events and provide easy access for trail users, thereby reducing the impacts of these visitors to the historic core. Though planned for a later phase, an option to move the easement road arose during site analysis to use an abandoned road alignment along the western edge of the property.

3. Increase Educational and Event Programming

A desire to increase educational and event programming came from the community and the park's stewards. However, there was a strong desire to minimize increased use impacts to the cultural and natural resources that draw people here. Wedding and special event sites were confirmed and set aside and various levels of infrastructure were woven into the site's fabric to ensure a balance between ease of use and protection of resources. The Barn and Tomseth House will remain as major educational and rental spaces for indoor, but will be supplemented by a new education and event space that overlooks the front entry meadow and serves as pivot point between the entry parking lot and the historic orchards and trails.

4. Preserve Rare and Prized Natural Habitats

Oak, prairie and riparian habitats were once important cultural, as well as natural, ecosystems. Native Americans maintained oak and prairie sites through large burns throughout the Willamette Valley. Today those practices are minimized and the habitat they created has nearly disappeared. Protection is only the first step in preserving these important natural systems. Targeted management techniques are even more critical to their sustained health and regeneration. This plan secured the protection of large swaths of oak savanna, prairie, woodland and riparian areas, but also worked with management staff and outlined various levels of cyclical maintenance techniques that are needed to ensure the continued health and vibrancy of these rare and prized habitats.

居住区

Residen

临水而居 —— 艾菲拉姆岛

Live beside Water — Ephraim Island

撰文 / 图片提供：EDAW Australia　　翻译：沈翀

该项目是位于昆士兰州黄金海岸北部的一处颇具名望的住宅社区，它很好地诠释了精巧与责任这两种景观设计理念是可以共存的。

2002 年，EDAW 的布里斯班分公司赢得了对艾菲拉姆岛的景观设计权。当地壮丽的景色促使了"沙丘"设计理念的产生，并使其成为了该项目的营销亮点。

在业已完工的一期工程中，设计师"打造舒适的住区外环境"的决定俨然已经成为当代海岸住宅景观设计的典范。

该项目的成功设计来源于对其周边环境的理解和尊重。景观设计必须要经得住强烈的日照、海浪的侵袭、高度的裸露以及海岸风和供水量少等问题的考验。预先做好准备则意味着居民可以在一个草木繁茂的岛屿环境下享受生活，而日后这一切也会在海岸环境下变得更加美好。

植栽

设计师从工程早期的总体规划到竣工期间的参与保证了设计初期的"沙丘"理念被贯穿实施到各个层面。设计师根据植物的功能特性遴选出几种植物，以求能经得住高度裸露的考验。

迄今种植的树木约有 62 000 株，其中大部分都是昆士兰州东南部本土的树种。将磨成细沙的贝壳粉用做护根，为植物创造了一种类似于其原生长环境的表层，因此所有植物的生长都很繁茂。

水的使用

避免将饮用水使用到景观各元素之中是该项目的核心目标，特别安装一个 100 000 L 的水箱用于贮存可循环利用的水。卡车将"乙级水质"的水运至岛上，并就地净化使其达到"甲级水质"的标准。该系统的使用将会保证岛上所有植物的浇灌都是采用可循环利用的水。

生态系统

在六幢公寓大楼中俯瞰到的中央湖泊就是从宽阔的水域直接引水的，给人留下的深刻印象不亚于在岛屿入口处桥上观看到的水墙。湖水被循环利用，自人工湖修建的一年来，这里已经成为了生机盎然的生态系统，海葵、海螺、海菜、鱼、虾等已在这里安家，蚌和其他很多海洋生物也已适应在水湾的潮汐中生活。从公寓中、栈桥上、公园中的阶梯剧场等有利方位观察，人们可以从潮起潮落中感受时间的流逝。

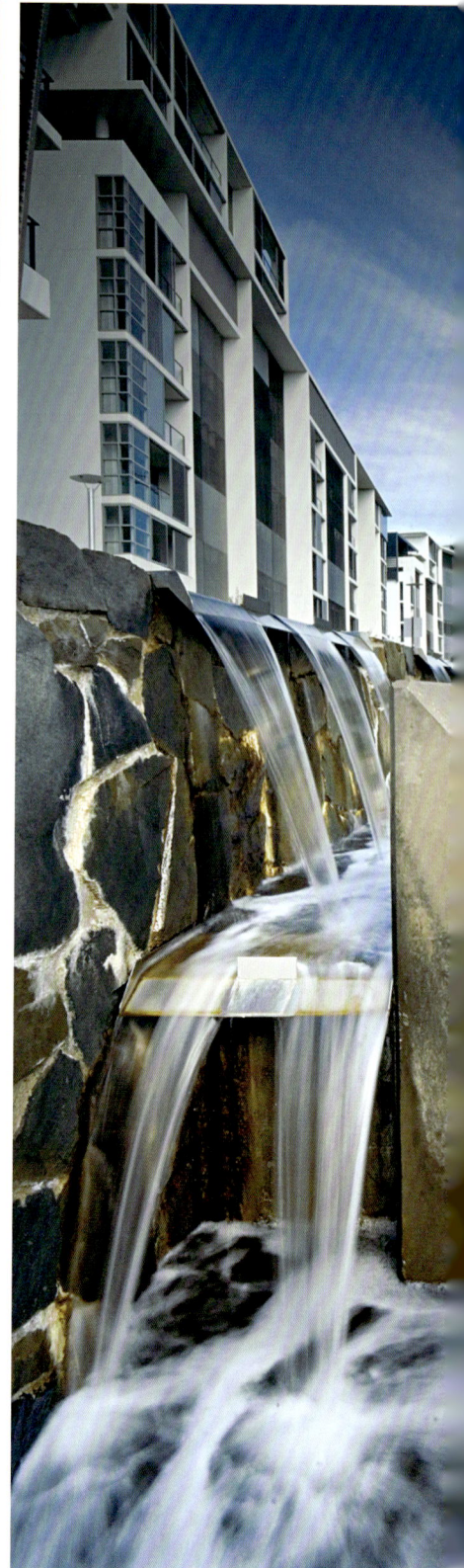

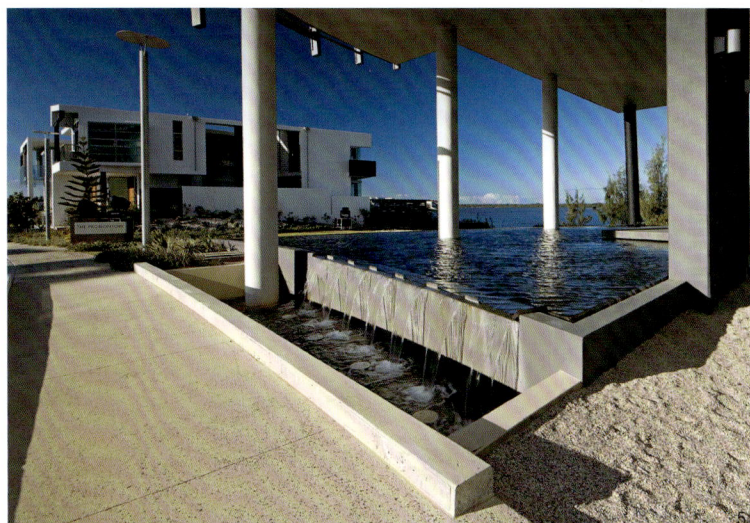

1 岛内的小瀑布之一
2 水面上的人行步道
3 水体提供一种安宁
  的听觉和视觉环境
4、5 住宅实例
6 滨水游泳池

Ephraim Island, a prestigious residential community at the north of Queensland's Gold Coast, shows that sophisticated and responsible landscape design can co-exist.

In 2002, EDAW's Brisbane office, won a competition to design the landscape at Ephraim Island with a dune concept inspired by the spectacular location, which has gone on to be the distinct marketing image of the project.

With Stage 1 already constructed it is clear that EDAW's decision to go "outside the comfort zone" of residential estate design has produced a new benchmark in contemporary coastal landscape architecture.

The success of Ephraim Island has come from understanding and respecting the site conditions. The landscape design had to withstand hot sun, salt spray, high exposure, off-shore breezes and require minimal watering. Doing the homework up front has meant that residents now enjoy a lush island setting which will continue to evolve and flourish in its coastal environment.

Planting

EDAW's involvement in the project from the early master planning stages through to completion has ensured that the initial dune concept has been carried through all levels of detail. Plants have been selected for their ability to withstand the highly exposed conditions on the island.

To date some 62,000 plants have been planted, with most being native to Australia and South-East Queensland. Using crushed shell grit as mulch has created a reflective surface similar to their natural habitats, and all plants are thriving.

Water

Avoiding the use of the mains drinking water supply for landscape elements was a core goal of the design. A 100,000 litre water tank has been installed to store recycled water. "B Grade" water will be trucked onto the island and then filtered on site to achieve "A Grade". This system will enable all the plants on the island to be watered using recycled water.

Living Ecosystem

The central perched lake, which will eventually be overlooked by six of the apartment buildings, is filled directly from the Broadwater, as is the impressive waterwall viewed across the water from the island entry bridge. The water is constantly recirculating and, in the twelve months since installation the lake has become home to a thriving ecosystem of sea anemones, marine snails, sea-green mussels, fish, and prawns. Mussels and other marine life has also quickly adapted to life on the stepped tidal sculpture in the main inlet. From vantage points in their homes, on the jetty, and at the amphitheatre park, residents can observe the passing of time as the tideline gently rises and falls against these markers.

1 葱郁的绿色空间环绕着居住区
2 岛内的生活格调缩影
3 水岸细部
4 朝向开敞水面的露天设施

蓝色住宅 —— 非洲大西洋海岸的山顶住宅

Casa Azul — Mountain Estate in Atlantic Africa

撰文：Emporis 网络编辑 Pedro F Marcelino　　　图片提供：Emporis 网络编辑 Pedro F Marcelino　Éric Loiseau　　　翻译：刘建明

1　住宅侧面图
2　主入口用红地毯来迎接尊贵的客人
3　房屋被一堵墙巧妙地一分为二，成为两处截然不同的空间——内部平台和仙人掌花园
4　主庭院延伸至水池庭院的部分

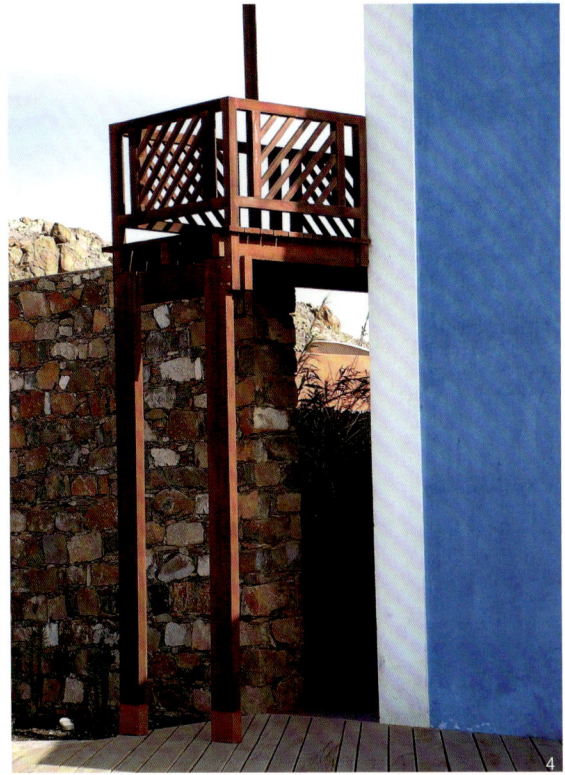

位于佛得角群岛的明德卢崛起于19世纪初期，这个港口城市当时主要作为横跨大西洋的船只补货交易的停靠点。虽然后来被并入葡萄牙殖民帝国，但却是英国人于1863年成功地在这个干燥的岛屿上站稳脚跟，并凭借其天然的煤炭资源和成立电报公司牢牢地抓住了本地的经济命脉。

无论是修建之初还是现在，19世纪70年代兴建的 Casa Azul（葡萄牙语，意为"蓝色的房子"）始终处于这个城市的边缘。作为英国公司管理者周末的度假别墅，这个地区从未发展成为主要居住区，最主要的原因可能是因为从明德卢市区到 Casa Azul 之间的路途必须完全依赖马车，并要在漫长而且尘土飞扬的道路上行驶。

这片小区的景观和文化背景引起了 Éric Laure 和 Laurent Loiseau 的关注。2001年，Loiseaus 家族决定拆除现有的建筑结构，以彻底摆脱旧式建筑的影子，内部空间被重新构筑，以实现多样化的庭院结构。灵感受到修道院回廊的设计启发，现存的房屋被一道外墙一分为二，成为各不相同的两处庭院，阶梯式设计的前院也具有岛上颇为典型的建筑风格。

砌墙用的水泥石块都是现场制作，本地石匠用独有的技术来堆砌传统的外墙，堆砌的材料是取自岛上的土块与小石块的粘合物。俨然乡村风貌的内里却隐藏着精巧工艺修饰后的现代水泥建筑，将本地的原材料和工艺与顶尖的现代建筑完美结合，带来了无与伦比的美学和建筑学效果。房屋和花园的建造同步进行，并尽量贴近自然。Éric Loiseau 并没有对整个场地进行细致的规划，某些部分仍保持原状，他相信假以时日，当小区的设计初具规模时剩余的部分自然而然地就会体现出最符合其自身特点的设计风格。每一项工艺的制作灵感都源于传统的做法，为了加固特别添加了铁氧化混合物，所有的地板表面都显现出勃艮第葡萄酒的颜色；在修道院回廊墙壁上的石灰中同样也添加了铁氧化混合物；添加了钴蓝色颜料的石灰被用来粉刷房屋外部的标志性墙壁；庭院里所有的木料表面都涂有 Laurent Loiseau 发明的亚麻油和铁氧化混合物。每个创意决策之间都有着逻辑联系，每个细节都展现出巧夺天工的绝妙，与其说是一个简单的投资行为还不如说是一部注入激情的心血之作。

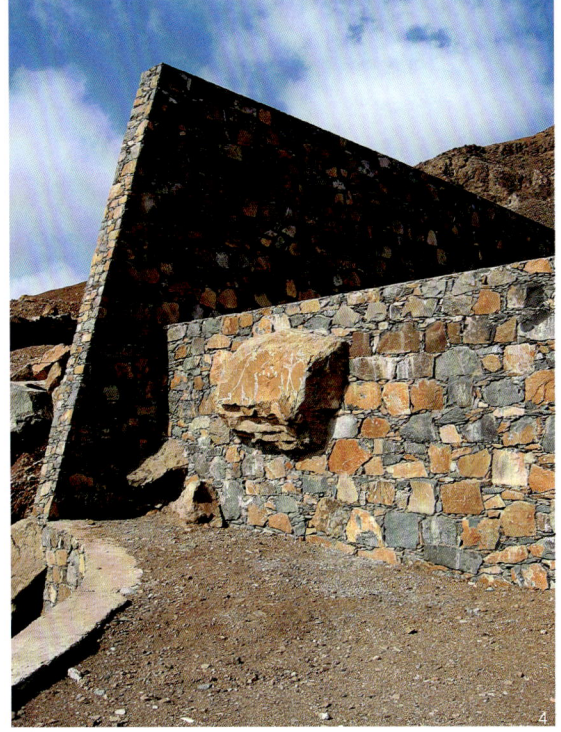

1　水池庭院与平台的全景

2　住宅区入口处仍延续着红毯这一主题

3　水池庭院——雕刻大量的岩石来装饰水池庭院的造型

4　墙壁将已存在的巨石镶嵌在内

5　水池庭院中的仙人掌花园

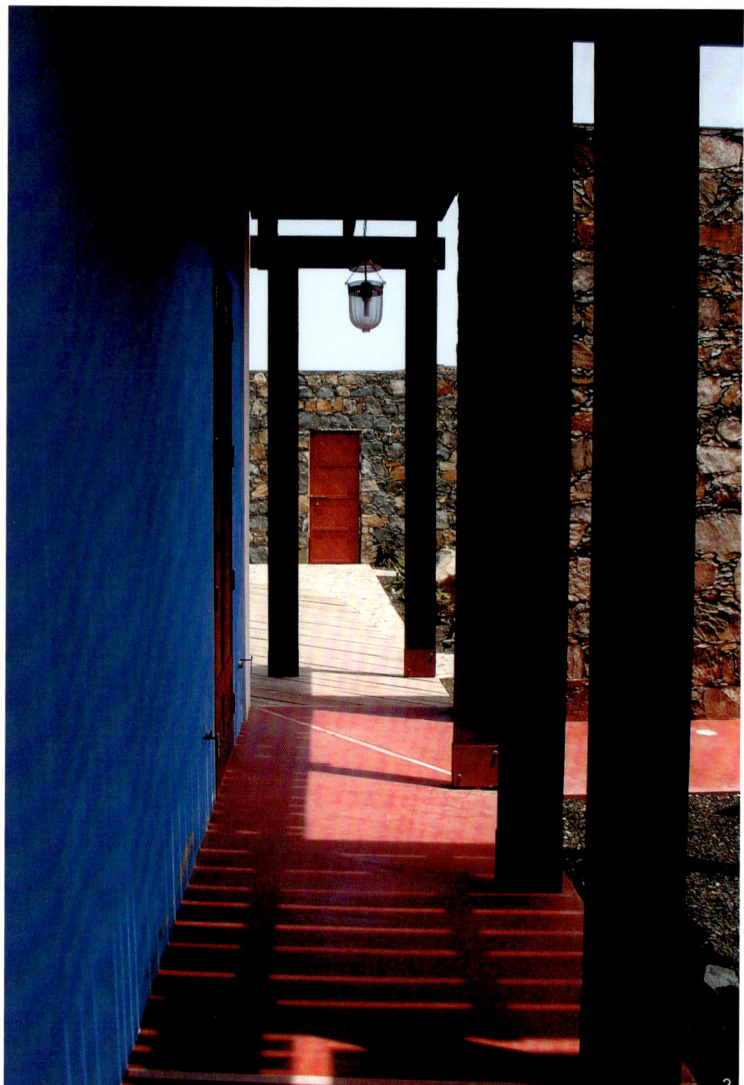

小区所在的地理位置人烟稀少，光秃秃的山地景致刚好符合这一基调。整个小区整日都暴露在阳光的照射下，无法不令人联想起 Frank Lloyd Wright 家乡亚里桑那州的 Taliesin 山。

乡土植被与竹子、无花果树、龙舌兰、阿拉伯树胶、仙人掌和干燥气候下山区特有的植物等混合栽植在一起，前院和内院的植被选择同样突出了这一种植风格。一些小型植被覆盖了部分土壤，可以起到护坡的作用。

公园内所有的区域都铺上了一层尖利、闪亮、黝黑的火山沙，以防止强风的侵袭。风成为了景观规划时所面临的最大挑战，但这不是惟一的挑战。面对强烈的、随时都可能出现的东风、令人难以忍受的热浪和少得可怜的灌溉设施，只有生命力顽强的植物才能在如此恶劣的环境中生存下来。

无论是房屋还是景观规划的区域都不得不考虑到已经存在的巨石，即便是这些巨石的位置影响到了景观设计，比如说水池的底部就是一块巨大的石头，许多工序都是在这块巨石上进行的。再次受到自然环境的启发，Loiseaus 利用这里的地势设计出中央的一处小型但却引人注目的坐席。

褐色的土坡、赭色的日落，Casa Azul 已经成为海湾中一道令人瞩目的风景线。在如此迷人的景色中，被当地人称之为"英格兰豪宅"的 Casa Azul 也成为了佛得角群岛的首家精品旅馆。

1　主入口红色大门敞开、红毯铺地，指明了入口路径

2　侧花园第一层门廊

3　受修道院回廊的灵感启发的主庭院

4　观望龙舌兰花园

5　以小径和早已存在的巨石为界，创建一个侧面花园

6　从后面的巨石分布排列观看房屋

1　树池细部
2　传统的干燥石墙掩盖住了现代水泥的建筑结构
3　建造水池庭院的三种材料
4　许多幼小的植物需要保护，以抵御强劲的东风
5　仙人掌花园中的小径烘托出远山蜿蜒的轮廓
6　结构墙的边缘

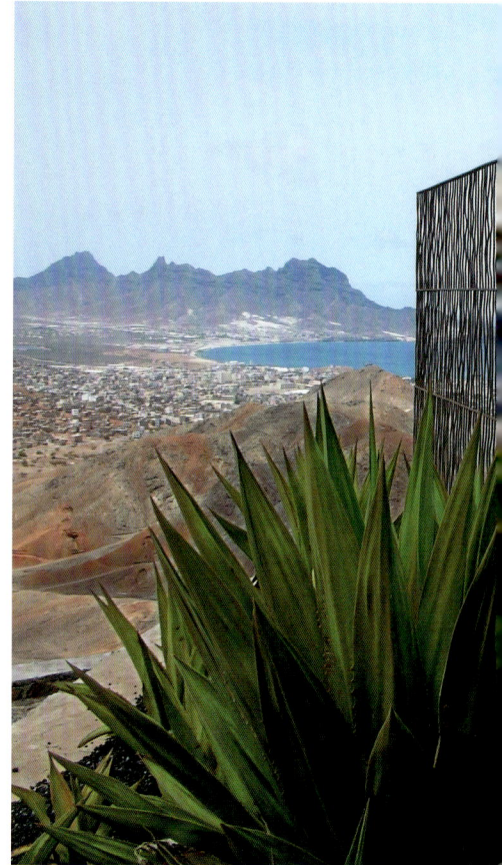

When the Loiseaus first traveled to São Vicente island from France, they had a simple holiday in mind. However, Mindelo, a charming bohemian city in West Africa for nearly 200 years, lured them with its warm people, warm music and warm nights. The family soon decided to relocate, not without traveling widely throughout the archipelago, in search of other potential locations. In October 2000, Laurent Loiseau—the son—was the first to settle in. One month later, a wrecked Casa Azul (Portuguese for "blue house") had been purchased and plans to revive it were underway. The harbour city of Mindelo was raised from scratch in the early 19th century as a re-stocking pit stop for transatlantic ships. Despite integrated in the Portuguese colonial empire, it were the British who first successfully settled in the dry island, around 1863, and from then on dominated the local economy, deeply dependent on their coal and telegraph companies.

Casa Azul, in the outskirts of the city then as now, was built in the 1870s as a weekend getaway for the administrators of English companies and never actually inhabited as principal residence, namely because the travel to and from downtown Mindelo had to be done on horse-drawn carriages, up the bulky and dusty mountain trails. Its location at 164m ensured the most privileged view of the Grand Harbour and the canal between Mindelo and the nearest island. It also metaphorically underlined the physical and social distance between the British and the locals.

The landscape and cultural background of this estate were what caught Éric, Laure and Laurent Loiseau's attention. In 2001, the family had concluded that the existing structure could not be kept, and demolition work started. They would finally move in in August 2004, although finishing touches would only be closed in April 2006.

Nothing was left to chance in this project. Construction economist by profession, and with a strong architectural background, Éric Loiseau—the father—applied his skills to the smallest details, dubbed in this work by his son. The original structure was modified to exclude obsolete constructions in the property, and re-organize internal space to permit multiple yards. The house currently has two cloister-inspired courtyards divided by an external wall creating two different spaces, as well as a laddered front yard in the typical style of the island.

Cement blocks for the walls were manufactured on site, and local stonemasons brought in the necessary know-how to create traditional external walls made of ocher blocks extracted in the island and kept together by smaller stones. Albeit rustic looking, they hide modern cement structures

1　地势较低的外庭被用做冥思场所

2　地势较低的外庭中夕阳下婆娑的树影

3、4　从日出到日落，阳光时刻眷顾着小区，并营造出许多有趣的影像景观

5　整体色调——钴蓝色和波尔多葡萄酒的颜色

6　带有精致典雅楼梯的内庭

7　从 Casa Azul 可以看到文森特岛明德卢海湾的落日景象

8　水池庭院里的小径

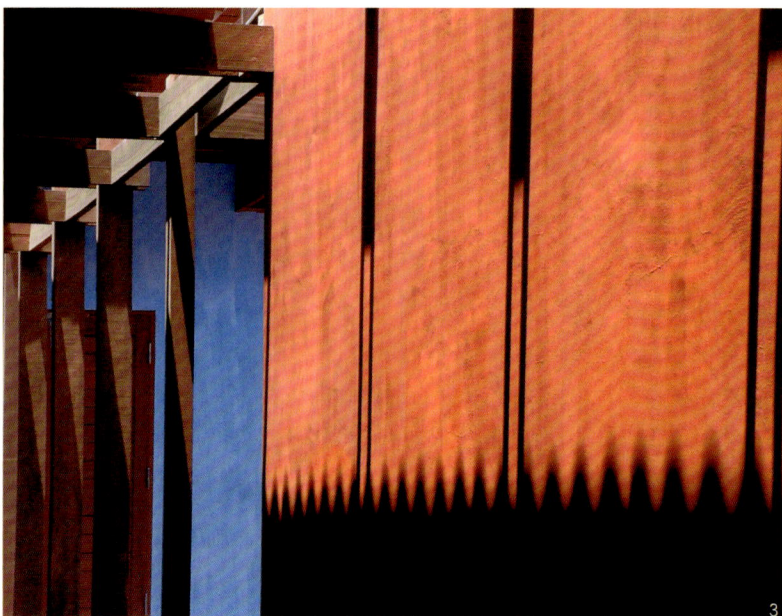

using refined techniques—like elsewhere in the property, local raw materials and expertise were combined with the best of modern processes, offering brilliant aesthetic and structural results. The house and the gardens evolved naturally. Eric Loiseau did not plan thoroughly, rather remained open to the answers that the property itself offered in due time. Each technique was thus inspired by the previous: to cement he added iron oxide pigments, creating the dark burgundy color that covers all the external flooring; the same iron oxide was then added to lime for the cloister's walls; cobalt blue pigments with lime was used to paint the emblematic external walls of the house; and all the wood in the yards was dyed with a linseed oil and iron oxide mixture invented by Laurent Loiseau himself. There is a logical chain between each decision, a personal touch put in to every detail that makes it a work of passion rather than a simple investment or architectural and landscaping project.

The landscape of desolated, dry mountains where the property is located matches its color palette. Exposed to all the sunlight variations, the estate at times evokes Luís Barragan's works or Frank Lloyd Wright's Taliesin, in Arizona. Combining endemic species with bamboo, Indian fig trees, aloë, agaves, acacias, desert cactuses and dry mountain flora, the choice of plants for the front and inner yards also reflect this. Some small plants cover part of the soil and help stabilize the slope, whereas all the garden area has been layered with sharp, shiny black volcanic sand extracted from nearby Calhau volcano, and used to prevent wind erosion. Wind was in fact the biggest challenge in the garden planning, but not the single one. Subject to strong, ever present east winds, only resilient plants could be used in order to resist them, the oppressing heat and poor watering. They were planted with a frequency that mirrors that of the mountain, although landscaping problems remain unsolved.

Both the house and the landscaped areas respected existing boulders, even when these seem to have been specifically displaced for a dramatic effect. The bottom of the swimming pool, for instance, is one single, massive, impermeable boulder, where access steps were carved in. In one location of the estate, just outside the house's walls, natural stone formations gather in a distribution similar to that of a Greek auditorium, having an inclined tree as the centerpiece of this "stage". Once again inspired by natural conditions, the Loiseaus plan to use this layout to sculpt a small but striking open air auditorium for small events in the middle of the mountain.

Set in a fascinating landscape, Casa Azul—locally known as Englishman's House or White Sand house—becomes the first boutique hotel in the Cape Verdian archipelago, clearly and tastefully influenced by the work of several contemporary architects and landscapers, as well as in local wisdom and traditions, fittingly integrated in the brown slopes and ocher sunsets on the bay.

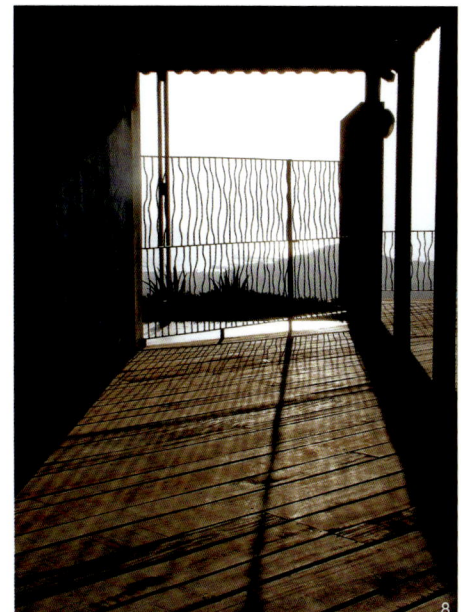

沐浴在朝霞与夕阳之中 —— 具有地中海风情的Bassil住宅

Sunrise Skiing and Sunset Bathing — The Mediterranean Flair of Bassil Mountain Escape

撰文：Pedro F Marcelino　　图片提供：Geraldine Bruneel/VDLA　Victor Gedeon (mount-lebanon.com)

1　平静的水面及云雾缭绕的山顶
2　静谧的环境让人们可以充分沉思和放松

该项目极具诗意和创造力。设计师运用各种空间元素、利用周边环境营造出一处宁静恬淡的居住空间。设计师非常注重该项目的整体协调，对设计的把握也是恰到好处。

——2008 年 ASLA 专业奖项评委会评语

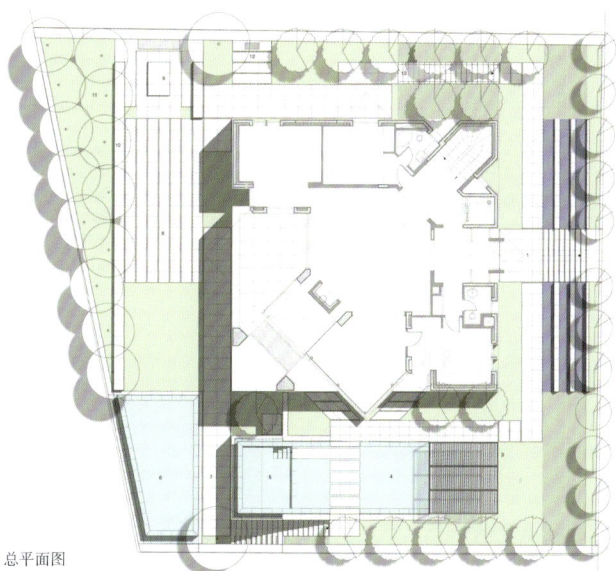

总平面图

1	主入口	5	悬壁屋顶	9	带有火炉的房间	
2	阳台	6	游泳池	10	线形长椅	
3	凉亭	7	酒吧	11	桦树林	
4	水池	8	休闲区	12	烧烤区	
				13	通往上层的楼梯	

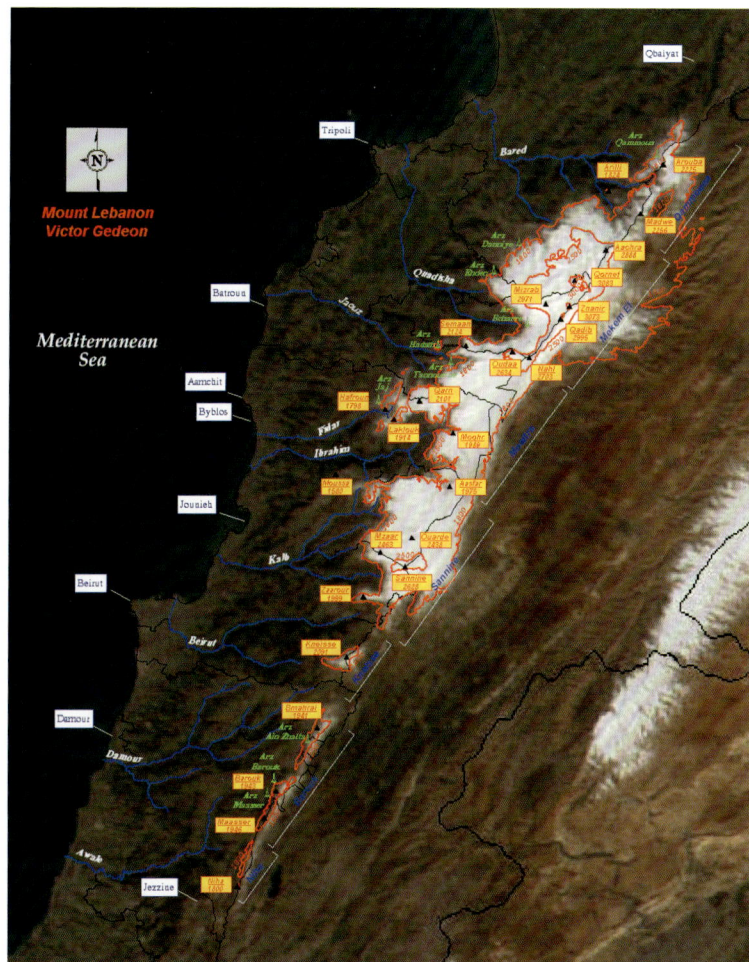

黎巴嫩山脉的地图

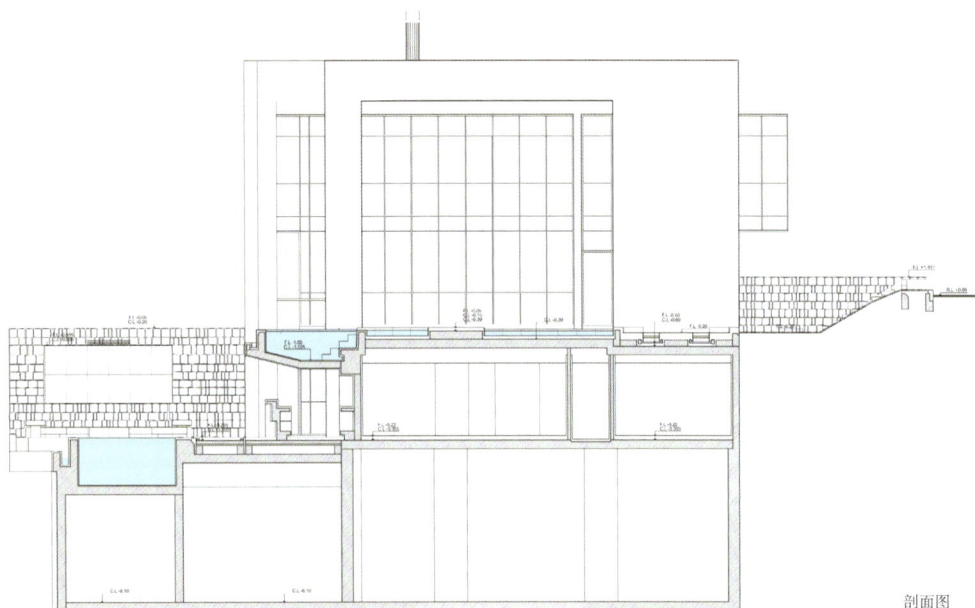

剖面图

地中海沿岸美丽的风景、黎巴嫩山顶挺拔的雪松以及贝鲁特的建筑风格，这些独有的魅力营造出该项目静谧与祥和的氛围。在历经长期的动乱后，黎巴嫩终于向世人展示其极高的美学价值和自然风光。作为中东地区不同文化与宗教的聚集地，这个国家已摆脱了战火硝烟及边境冲突问题，逐步走向和平。每一次灾难过后，黎巴嫩都会变得更加坚韧与顽强。作为黎巴嫩的首都贝鲁特仍同往常一样繁荣发展着，众多的设计师以及艺术家齐集于此，使许多欧洲小城都显得黯然失色。

1　树木掩映下的无边界泳池
2　连接室内与户外花园的走廊
3　水池夜景

户外空间手绘图

曾几何时，贝鲁特被誉为是"中东地区的巴黎"，其精致的餐馆、浓厚的文化氛围以及在城市中流行的新浪漫主义风尚都会使人联想起巴黎。在迪拜、卡塔尔、巴林等城市崛起之前，很多富有的中东贵族都来这里享受都市生活。如今，被取代的是混凝土的碉堡、弹药库以及瞭望塔。经过战争的洗礼，更加坚定了黎巴嫩人民追求未来生活的信心。最具代表的便是设计师 Sadar Vuga 和 Alvaro de Siza 在废墟上重建的建筑，以其独特的建筑语汇与人们进行交流——醒目的粗线条，同时也体现出建筑的亲和力、创新性以及纯净感。

Bassil 住宅的景观设计师 Vladimir Djurovic 出生在贝鲁特，是这一地区颇具影响力的景观设计师之一，曾荣获 Aga Khan 建筑师奖、国际设计奖以及都市风景建筑奖。在他的眼中，黎巴嫩的一年里有 9 个月的时间可以居住在户外，室外空间和室内空间同等重要。于是，将室内外完美融为一体的 Bassil 住宅便由此诞生了。

户外休憩空间手绘图

1   火炉为夜晚还在户外的人们带来了一丝温暖
2   泳池夜景

Bassil 住宅位于黎巴嫩 Faqra 市的一个度假区内，距贝鲁特仅一个半小时的车程，是一个多功能的私人住宅，也是炎热夏季里一处清凉的避暑之地。考虑到场地有限的空间、4.5m 高的建筑以及崎岖不平的山体，设计师调动了一切元素营造出柔和、静谧、舒适的意境以满足主人的需求。露天浴缸、户外酒吧、游泳池、设有长椅的阳台、带有火炉的房间以及装修精致的烧烤区，每一样都能够带来不同的体验。

从入口一直到住宅前面，都种植着清香的薰衣草；玻璃镶嵌的餐厅对面是一个凹进去的、带有顶棚的起居室；犹如镜子一样平静的水池反射出气势磅礴的山脉，山顶时而被积雪覆盖、时而在浓雾或云层的掩映下若隐若现；水池被石板小径分成两部分，将客人引入由石材和木材制成的酒吧区；游泳池为这里秀美的景色锦上添花，带给人们无限的遐想；设有长椅的阳台为人们提供了一处休憩空间；烧烤区位于落满枫叶的木屋之中，与这里的景色完美地融合。

建筑与各种户外的休闲空间完美融合，共同构成了这处集高山魅力与地中海特色于一身的秀美景观。

The beautifully accented arid landscape of the country's Mediterranean coast, the plentiful cedar forests atop Mount Lebanon's snowy peaks, and Beirut's legendary architectural charm and innovation suggest a peaceful enjoyment that is not always there. For all its aesthetic value and natural highlights, Lebanon has been through more troubled decades than any country deserves. An intercultural and inter-religious hub in the Middle East, this small country has been through civil war and enough cross-border conflicts to lower its arms. Instead, Lebanese seem to stand proud after each hiccup. Beirut is as thriving as ever, with a boisterous nightlife and enough architects, designers and artsy venues to make most of Europe's small capitals blush. As this article goes to print, a new armed conflict looms in South

1　水面倒映出褐色的山体
2　无边界泳池上的石板，犹如漂浮在空气中
3　休闲区的景色

Lebanon. Yet, once again, the Lebanese people's attitude is one of stoic optimism, certain that all parties will pull back at the last minute—enough is enough, and the whole country seems to know.

There was a time when Beirut was dubbed the "Paris of the Middle East", referring to the sophisticated café culture, high culture and neo-romantic façades reminiscent of the French capital. Before Dubai, Qatar and Bahrain had even popped out of the desert, well-heeled Middle Eastern families flocked in for the capital's social life. Much of that city has now long vanished, the urban landscape instead peppered by concrete bunkers, ammunition depots and watchtowers, urban voids and open wound in the urban fabric. But also this has been re-routed by resilient Lebanese creators, steadfast in pushing life forward despite the traumatic experience of civil war. Some of the most recent architecture builds on existing war time structures, or uses an architectural language that communicates with them—harsh, aggressive lines that simultaneously feature the graciousness, innovation and purity of Sadar Vuga or Alvaro de Siza's work.

Born in Beirut, Serbian-Lebanese Vladimir Djurovic is one of the region's most respected landscape architects. After a long period abroad, which included a stint at Edaw in Atlanta, Djurovic returned to the city in 1995. His practice has grown ever since, gathering awards such as

凉棚及楼梯手绘图

the Aga Khan Award for Architecture, the International Design Awards or Cityscape Architectural Review Awards. Djurovic's work unmistakably relates back to the brown Mediterranean landscapes, colours and scents. In a land where (in his words) "you can live outdoors for nine months a year, where outside spaces are as important as the interiors", it is thus hardly surprising to come across a project such as Bassil Mountain Escape, with a conceptual clarity that tests boundaries, and an uncompromising fusion of indoor and outdoor living.

Only one and a half hour (upward) drive from Beirut, Bassil Mountain Escape is located in the Faqra Club estate, a development within the Swiss-style mountain resort town of Faqra, and nestled between the reputed Qanat Backich and Ouyon es Siman skiing ranges in Mount Lebanon and the Keservan Mountains, respectively. The residence, privately owned by Jimmy Bassil, was commissioned as a holiday house, and projected with sole focae in contemplation, relaxation and entertainment. The landscaped area surrounds the chalet, spanning from simple shrubs garnishing slab paths in the narrowest sections of the site, to the wider, but still very limited spaces of the back.

Constrained by the relatively small size of the plot, the dismal 4.5m construction setback and the rugged topography of the mountain, the firm used an encyclopedia worth of illusions to yield a sense of peace, isolation and expanse in all the spaces of the property, catering to every possible social and family need.

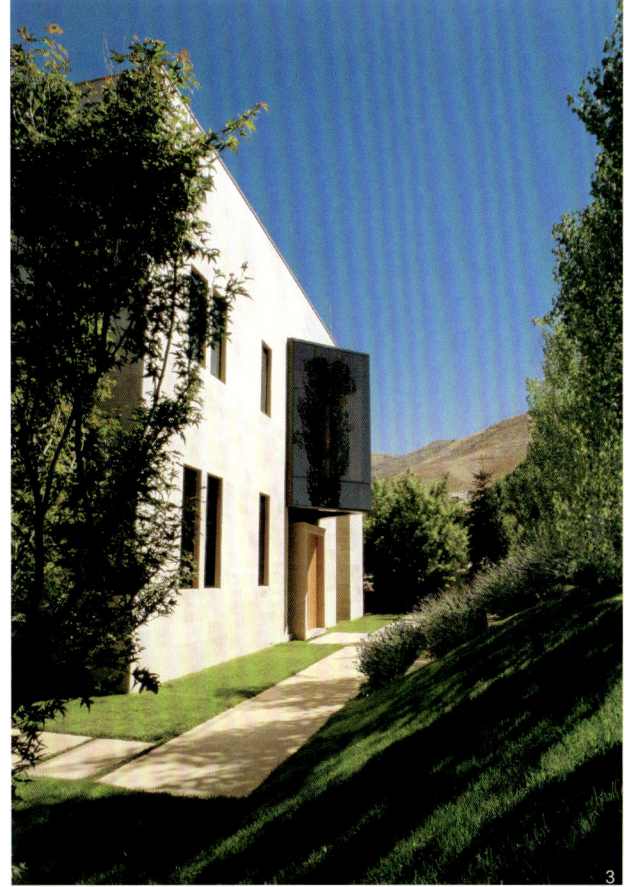

The result is a series of sitting areas with different atmospheres, affording multiple entertaining programmes and experiences. A water mirror in one of the corners of the property provides a stunning view of the sharp peaks nearby. In the outer end, a cantilevered jacuzzi overlooking the clouds doubles as roof to an outdoor bar, a swimming pool, a large entertainment terrace with a long linear bench, an elegant fire place in the wall and a beautifully crafted BBQ area.

In the upper level, the extremely narrow strip in the front of the house prompted an outside-the-box design solution. The entrance is marked by a solid, descending (rather than ascending) stone staircase framed by perpendicular lavender-filled planters running along the length of the house. Facing the glass-clad dining room, a recessed sitting area shaded by a canopy, a raised water mirror, and a cantilevered negative-edge jacuzzi align to capture the breathtaking view of the mountains, at times covered by snow, at times invisible behind thick fog or low clouds. The water surface is divided in two by a floating zebra crossing leading guests down to the warmer bar area built with solid local stone and red cedar wood.

In this level, the infinity pool enhances the outstanding panorama from both the indoor lounge area and from the outdoor entertainment terrace and sitting area, suggesting a vast space that is indeed not within the property. The long linear bench doubles as a safety balustrade and resting area for large gatherings. The BBQ zone is concealed in a wooden shelf between maple leaves, seamlessly fitting in the generally clean lines. Guests are lead back to the entrance via a narrow staircase embedded in the ground.

Overall, the chalet, outdoors entertaining and chill-out areas result in the ultimate mountain refuge, a hybrid between Alpine glamour and Mediterranean pizzazz.

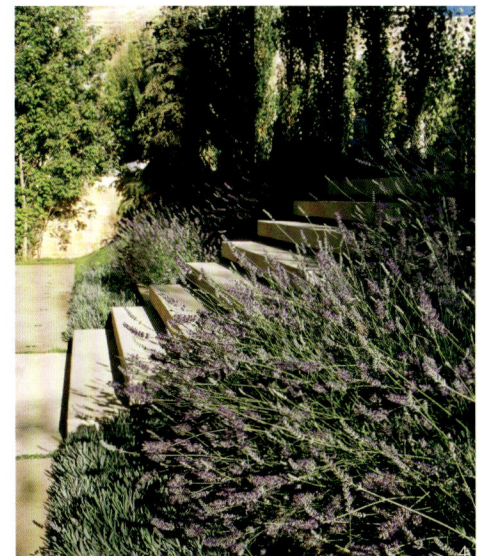

Grantoften住宅区景观设计

The Landscape Design of Grantoften Tower Buildings

撰文 / 图片提供：Lone Van Dears Tegestue   翻译：沈翀

1 红色建筑
2 平坦地面上的种植槽与斜坡相得益彰

哥本哈根市中心以西20km有一大片兴建于1968年～1969年的住宅区，包括可以租住的公寓和联排房屋。该区共有1340套住房，但最具特色的应数三栋内部设有836套房间的8层建筑。

20世纪60年代末，随着理性建筑的概念逐渐被人们所采纳，而营造出绿色环境。笔直的道路、和谐的植栽、低矮的灌木、高大挺拔的树木、灰色混凝土铺筑的人行道和车行道等，整个城区的全貌大致如此。

2004年，住宅区被重新翻新，三栋塔式高楼以南的环境也得到改善。由于原有植被的维护难度大、不具备观赏价值，所以必须将其清除。

步行道

步行道的设计灵感来源于西班牙道路设计，此处开放空间内少有装饰，可供游人散步或休憩。金黄色的地面铺装营造出一种温暖、明媚的气氛，并形成一个沉静的衔接平面。呈波浪形的黑色地砖贯穿于金黄色的地面铺装中，为整个区域增添了生机。

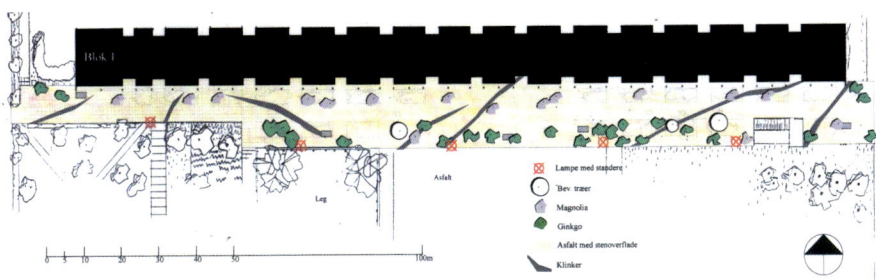

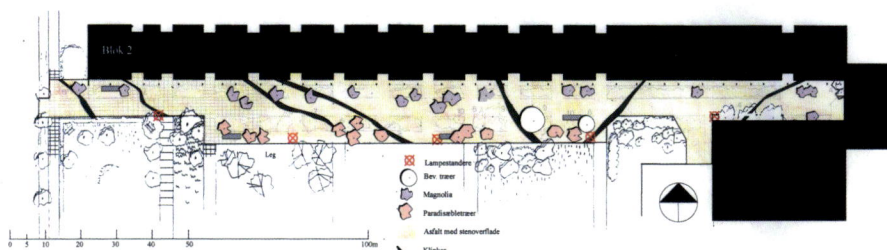

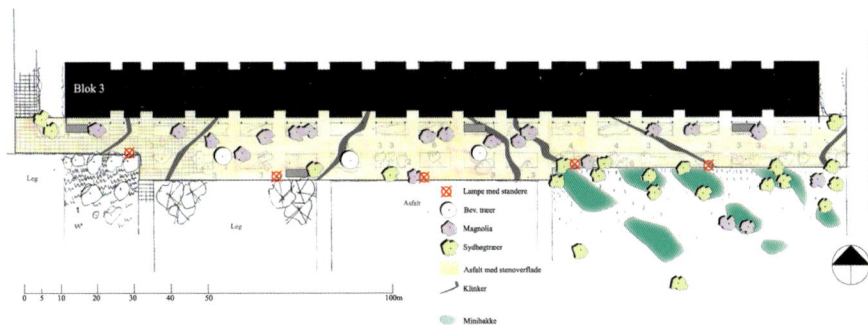

三座建筑平面图

植栽

　　设计师精心挑选了适合整个住区的植栽，种植在富于特色的种植槽中。种植槽的边缘采用钢板围合，形态各异，用橄榄形、圆形和三角形分别指代不同的建筑。

　　三座塔式高楼都建有混凝土廊柱，攀缘植物沿着廊柱上的钢丝绳生长，形成一个绿色垂面。

石块

　　设计师除了运用植物之外，也采用各种各样的石块作为装饰。灰色的中国花岗岩被分割成小块，2块～3块置于一起作为石凳。设计师还选用了几块巨石作为天然石雕，放置于步行道之上，颇具特色。每座建筑都有其特定的颜色——红色、绿色和蓝色，因此设计师选用了不同的巨石指代不同的建筑。在居民和景观设计师的共同筛选下，最终确定了令人兴奋而又形态各异的巨石。这些巨石的总重量大概在10t～20t之间，红色的是瑞典花岗岩，绿色的来自于挪威最北端的 Atlantis kvartsit，蓝色的是来自于加拿大的蓝晶石。这些巨石在阳光的照射下耀眼而夺目，在雨水中则五彩斑斓，给人以丰富的想像空间和体验。

West of Copenhagen, 20 kilometres of the city centre, a large housing area was built in 1968-69, with flats and terraced houses reserved as rental residences. The area consisted of 1340 residences all in all, but the most characteristic part was three eight stories buildings containing 836 flats.

Back in the late 60's, the green surroundings were rationally planned, by the use of cranes and as the idea of the rational architecture had become generally accepted. Long straight lines, uniform planting, low undergrowth, a few stemmed trees and traces for driving and walking made in grey concrete were the overall picture.

In 2004 a renewal of the housing was begun and a possibility to also renew the boring, worn-out surroundings south of the three tower blocks had been given. It was clear that the old planting was to be removed as it needed too much care and was without visual value. Also the existing covering was worn down with large holes in it.

The Rambla

As a superior theme the inspiration was found in the Spanish rambla, which is an open area that invites people to either stroll or take a break and is only scarcely furnished.

A golden yellow covering was the over-all sight along the facades. It was to create a warm and bright impression and form a calm joint surface. To apply life and movement to the area, wavy paths of black tiles were incorporated abeam of the golden longitudinal direction.

The Plants

Stemmed trees and flowering bushes were chosen in cooperation with the residents. Different plants for each of the three blocks were placed in low screening beds, constructed from rims of steel which surrounded the mould. The beds were made in three different shapes, the eye, the circle and the triangle, chosen to characterize each their building.

The three tower blocks had bearing constructions of concrete pillars. Up these pillars were placed steel cable with fast growing climbing plants which in time would show as green vertical surfaces.

The Stones

Apart from the plants, different kinds of stone were used as furnishings. Grey Chinese granite was cut into dices and placed in small groups, to be used as chairs. As a very special feature, large unhewn stones were selected and placed as sculptures on the ramblas. Each tower building had its own colour; red, green and blue and so the stones were chosen to match these colours.

A group of residents joint the landscape architect on a trip to a stone storage and selected these exciting and expressive stones. They weighed somewhere between 16 to 20 ton all in all. The red stones were Swedish granite, the green stones were Atlantis kvartsit from the northernmost part of Norway and the blue stones were cyanit quarried in Canada. The stones were transported to Grantoften and today they lay as elements that appeal to the imagination and offers many experiences, sparkling in the sun and deeply saturated with colour in the rain.

1 三角形种植槽
2 采用中国花岗岩制作的石凳
3 绿色建筑前的圆形种植槽
4 橄榄形种植槽
5 种植槽被随意地布置在广场上

景观与建筑的对话 —— 其中有象

Dialogue between Landscape and Architecture — Growing in the Voids

撰文：安庾心　　图片提供：安庾心　AECOM 中国区规划 + 设计（广州办公室）

该项目景观设计采用了与建筑立面相似的设计语言和手法，将 3D 空间的线形应用于植栽和花池之中，在叙述虚实关系的同时与建筑形成呼应。人性化尺度的线形作为一个统一的符号将较为散落的户外空间组织起来，营造出连续性的空间体验。自由生长的植栽与运用线性手法表达的景墙及花池形成强烈的对比，以此形成该项目鲜明的设计语言。设计师致力于将具有生命力的植栽及功能空间共同融入建筑中，形成不同的景观形态，尽显"其中有象"的意境。

该项目的设计有多种不同的景观形态且处于不同标高的户外空间，包括主入口街景、别墅道路景观、别墅间的小庭院空间、会所景观及其屋顶花园、高层平台花园及其屋顶花园。

主入口的大门被视为豪华高贵的象征，这个尺度的设置并不在于突显豪华，而是为了与主干道上的国际学校立面和谐一致。景观设计师对入口道路的轴线进行了偏移调整，由此形成了不断更新的景观体验。人们进入大门后要从两面水景墙之间的小径穿过，在潜意识里构成穿越过渡区，营造出"过桥"的概念。别墅区的道路采用无道牙的处理方式，统一了所有的户外空间；路面铺设的红色透水砖与别墅立面的部分颜色相互呼应，也与白色建筑立面形成对比。道路两旁阵列式种植的大乔木打破了别墅区原有的天际线，同时也强化了居住区的私密性；竖向生长的树种在打破单一天际线的同时，也让自然的生命力穿透进来——和谐中见睿智，秩序里显生机。

从空中可以俯瞰别墅区块状布列的绿色屋顶，设计师将自然的生命力带入项目中，使植物与建筑相辅相成。对于位于别墅之间的小庭院，设计师力求通过植栽的质感和颜色渲染，形成具有度假氛围的花园，植物景观构图由灌木及地被植物组成，在有限的空间中植入优良的品种，构成了风景如画的花园。垂直的植栽丰富了"虚"的空间，通过微妙的层次展现出景

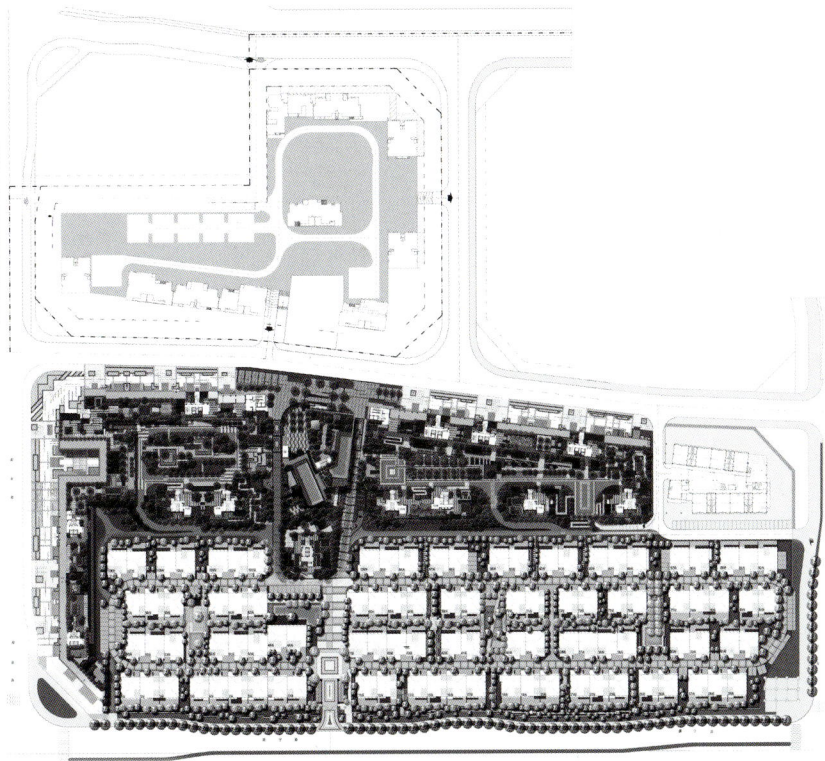

一期首层平面图

观的四季变化;"实"的建筑则有如白色的画布,而植栽相当于颜料——有机的植栽在"虚"的空间中生长,形成了栩栩如生的画面。多个功能灵活且使人放松的小空间营造出度假的氛围——椅子、景墙、花架这些景观小品作为统一的景观元素策略性地摆放,用于休憩与遮阳的同时也起到了分隔空间的作用,形成街道与内庭的过渡。小庭院空间是别墅区惟一的公共空间,人们可以在这里晒太阳、阅读或冥想。庭院内运用了不同形式的水景,如竖向水景墙和环形的浅水池等,并在设计时尽量控制好用水量,保证即使在没有水的情况下,也能形成完整的景观。

该项目最大的亮点是会所区的景观——两个泳池分别位于屋顶花园和地下广场的底层。其中,屋顶花园的泳池运用了玻璃盒的概念,即一个无边际的泳池,而到达该泳池的路径设计则强调了竖向动线的运用,具体体现在空中玻璃桥及户外楼梯的设计中;地下广场的泳池高出底层地面1.5m,兼备了水景墙的功能,到达该泳池需经过一段户外路径,人们可以在途中欣赏到不同角度的景观。屋顶利用植栽围合边缘,在保证了活动安全性的同时,也空闲出更多的中央空间;植栽的层次变化及灌木的搭配让屋顶花园呈现出构图式的景观。

高层平台花园在铺装、墙壁、植栽上统一运用了线性的设计语言,达成与场地的呼应。另外,入口瀑布式的水景加深了景观印象。玻璃电梯及残障坡道的设置满足了人们的通行要求,植栽在遮挡建筑立面的同时也柔化了建筑空间。屋顶花园位于平台花园之上,可经由户外台阶到达,设计师并没有赋予其固定的功能,因此这里可以作为自由空间用于户外锻炼及休闲娱乐,让人们可以在其中自由地开展各种活动,赋予该项目新的生机。

景观层次的强化与丰富在该项目的景观设计中起到了重要作用,同时也满足了人们内心对自然的需求。这种充满颜色和质感的自然洗礼让"虚"的空间随着时间的变化而不断净化、更新,这种变化发展是超越"实"与"虚"的概念的,在"虚"的空间里孕育出的生命力将不断成长、生生不息。

1 主入口大门及绿色屋顶
2 私家花园的泳池

Strong datum lines appear on the building facade as balconies. These three dimensional lines make a strong architecture statement about solid and void. Landscape design takes on a direct approach in responding to the architecture by using a similar design language. Landscape designer uses datum lines to connect the outdoor spaces and these three dimensional lines become the planter walls, feature walls as well as row of trees. By using these datum lines as a common design language throughout the development, subconsciously, these lines become the most memorable elements. In addition, planting design also provides a rich layer that is filled with color and textures. The juxtaposition between the rigid line and the organic vegetation becomes the most memorable experience in Kawloon Jade.

There are several types of outdoor spaces within the development: the main entry drive, the streetscape at the villa district, courtyard spaces between the villas, the club house area and its roof garden, as well as the high rise podium gardens and its roof garden. These spaces are located at different area within the development, and each space is at a different level.

Landscape designer adjusts the road alignment of the original layout in order to enrich the road experience as well as to avoid seeing the entire development right from the entry. People perceive the large gate way as a luxury icon for Kawloon Jade despite the size of the entry gate is a direct response to the surrounding context, i.e. the facade of the international school across from the entry gate. The entry is flanked by two water features walls, and a reflection pool. The entry design is intended to imitate the crossing over the bridge experience. It acts as a threshold dividing the new and the old. Next to the entry, there are rows of large canopy trees planted along the villa. The designer is intended to break the sky line of the buildings and adds privacy to the resident. The curb-less street unites the overall ground plane as whole, while the red concrete pavers gives a good contrast to the white facade of the villa. Trees selected are taller in height as oppose to its width. These trees are inserted between the villas to enrich the experience.

Bringing nature to the project is one of the main design objectives. Growing trees and shrubs in the void does

not only bring life to the project but also soften the solid. Planting begins at the roof top where all villas have green roofs covered with lawn, a deliberate attempt to make the uninhabitable roof pleasant to look at from above. Coming down to the ground plane, planting design is intended to create gardens that have a resort ambience and filled with textures and colors. The gardens composed of three to four layers of shrubs and groundcovers. Selectively, species are placed to compose a picturesque garden. These gardens enrich the void space and brought a subtle layer that truly speaks for the value of landscape architecture; bringing in the seasonal quality to the development. The solid provides an excellent backdrop for the color and texture. In order to achieve the resort ambience, several small gathering places are created within the villa district and they all have water features elements. Some are water feature walls, where others are shallow pools. All the water features are designed to minimize the water usage and designed to look great without water. The courtyards allow passive activities such as reading, strolling, and meditation. There are feature landscape elements used throughout the villa

1   屋顶无边际泳池
2   下沉会所泳池
3   穿插的景观空间
4   会所双泳池鸟瞰图

district; from trellis, bench to feature walls. These elements are strategically placed and they are well integrated within the design. Particularly, the trellis acts as shading device as well as divider that marks the transition between streetscape and inner courtyard.

The highlight of the project is located at the club house area. There are two swimming pools at the club house, one at the sunken level whereas the other one is on the roof. The pool at the roof garden is an edgeless pool, and it looks like a glass box. The roof garden is heavily planted with lush vegetation around the perimeter to keep people away from getting too close to the edge. Through planting with multiple layers of shrubs and ground covers, the picturesque garden is filled with colors and textures. The landscape design at the club house encourages vertical movement through the use of the sky-bridge and the outdoor stairs. The vertical movement is emphasized by creating a destination point, i.e. the swimming pool that is only accessible through the outdoor stair. The outdoor path offers multiple viewing angles for the sunken plaza, while walking down towards the pool. The sunken level, which is accessible only through the indoor lift and stair case, has the direct entry to the parking garage. At the Basement level, the pool is elevated 1.5 meter above the finish floor which makes it a great visual water feature element.

Both the West and East podium garden has the same design language; a site specific response to the context. Both gardens have taken on the form of a linear garden. Each podium addresses the entry condition as well as the ADA access by means of glass elevator, and a ramp. The water feature cascading down at the entry creates a resort ambience. Through planting, the garden has successfully grown to partially block the building elevations, thus soften the space in between the towers. There are several roof gardens on top of the podium which is accessible only through the outdoor stairs. These additional roof gardens are the additional outdoor rooms for exercise, reading as well as playground for children.

Landscape design enriches the overall experience within Kowloon Jade. Through planting design, the voids are connected both visually and mentally. With the consistent design language such as planter walls and feature walls, the overall development is coherent and it gives a strong impression of being a well thought out landscape design. Making this connection with nature is the main objective for the landscape architects and the concept of growing in the voids does not only make the local resident feel at ease of their own backyards but also metaphorically symbolize the ever changing nature and its possibility of growth.

纹饰图案的魅力 —— Paige住宅

Glamour of Filigree Pattern — Paige Residence

撰文：Roderick Wyllie 图片提供：Jeremy Harris 翻译：申为军

在2008年旧金山装饰展上，Surfacedesign Inc.(简称SDI)展示了一处创意十足的住宅入口区的景观设计。这幢位于旧金山的住宅建于20世纪初期，属于历史建筑。设计师采用大胆的设计手法，重新定义了从街道到建筑前的主入口区，从而强调出这幢标志性建筑与其所处的陡峭地形之间的关系。

设计师结合该建筑外立面20世纪早期的华丽风格，重新解读了铁艺栏杆上的透雕图案，并由此获得灵感，设计出两座倾斜的大型立体花园，并模仿建筑及铁艺栏杆上的纹饰在其表面覆土植草，刻画出令人惊叹的图案。倾斜的造型突出表现了该项目所在街区的特殊地形——旧金山最陡峭的街道之一。

立体花园的纹饰图案活泼有趣，一直延伸到入口台阶上，从视觉上将这两座立体花园联结在一起；四面不显眼的镜子映射出倾斜的大型立体花园、栏杆和天空，带来丰富多变的图案。

与建筑外立面丰富的纹饰相比，立体花园在材料和植被的选择上很简单，突显出简约的设计风格。花园中嵌入的可反射光线的金属制品与修剪过的草皮及深黑色的覆盖物形成鲜明的对比；花园中多年生的植物时常会打破图案的限定，使整体设计更加柔和；花园中原始的深红色砖墙被保留下来，以衬托浅绿色立体花园的新颖设计。

涡卷形照明设施再次呼应了铁艺栏杆的抽象图案。夜晚，灯光或投射在建筑外立面上，或被低处的镜子反射回来，形成了绝佳的照明效果，既标示出住宅入口，又给访客留下了深刻的印象。

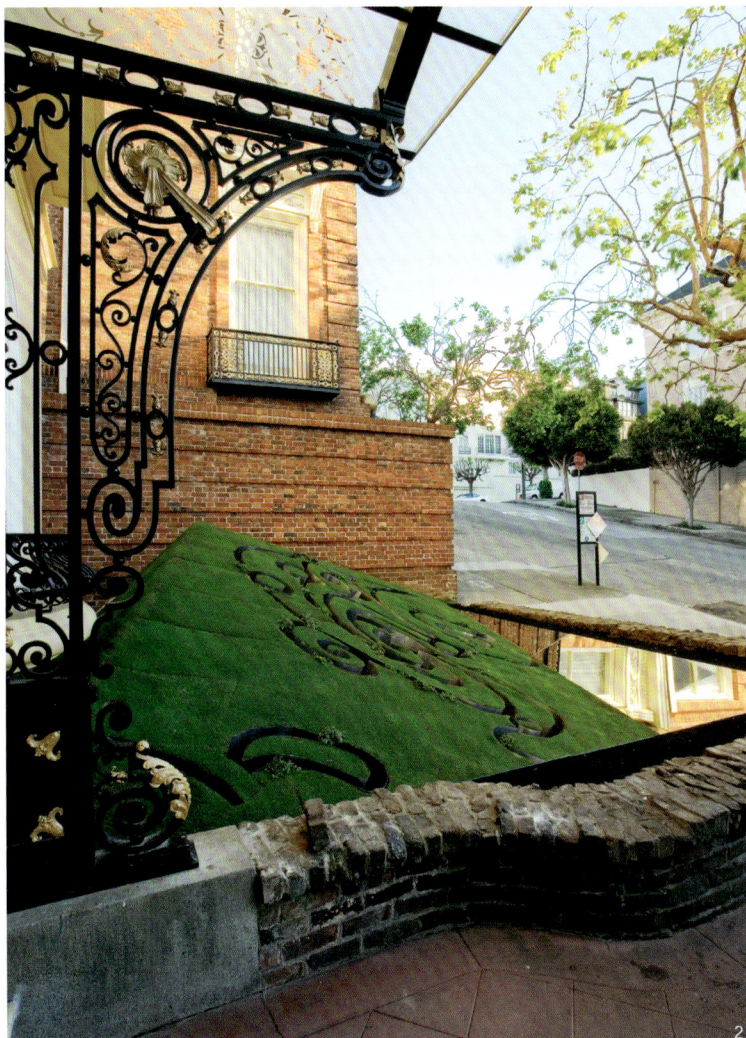

建筑草图

铁艺栏杆的图案

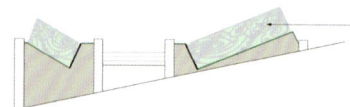

立体花园

场地剖面图

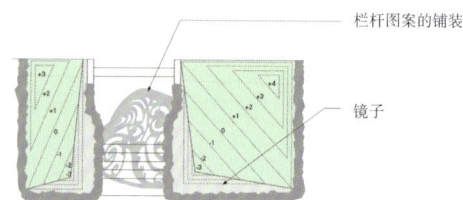

栏杆图案的铺装

镜子

坡度示意图

1　夜景

2　纹饰图案细部

3　镜子反射的建筑外观

Surfacedesign Inc. (SDI) design for the 2008 Decorator's Showcase creatively transforms the entrance landscape for an historic early 20th century San Francisco residence. SDI highlights the relationship between the iconic house and its steep hillside terrain by redefining the main approach from the street to front door in a few bold design moves.

SDI drew upon the house's opulent early 20th century façade and reinterpreted the filigree ironwork of the railings as an inspiration for a pattern inscribed on two large, tilted, earthen planes. These two steeply sloped grass parterre gardens highlight the extreme topography of one of San Francisco's steepest streets and bring an increasing awareness to the site's location.

The playful filigree pattern of the earthen planes extends from each garden and stretches across the entry steps, while visually connecting the two gardens on approach to the house. Four large hidden mirrors heighten the play of light and perspective by reflecting the gardens' topography, the house's railings, and the sky above.

The simple plant palette and material selection emphasizes the direct design approach in comparison with the opulent original design of the house. Reflective metalwork that forms the inlaid parterre garden contrasts both with the manicured short grass sod lawn and dark black mulch. Garden perennials intermittently break-up the larger filigree pattern and soften the overall design. SDI retains the original dark red clinker brick walls, which help to further highlight the newly sculpted light green parterre gardens.

At night, the design is illuminated with a scrolling light show, again, an abstract of the ironwork. The lighting effects are both reflected on the façade of the building and in the mirrors at the base of the parterre planes. The overall effect provides for a nighttime optical event that both illuminates the guest experience and showcases the house entry.

半岛听"风" —— 巴瑞洛切之屋

Listening to the Wind on the Peninsula — Bariloche House

撰文 / 图片提供：Jimena Martignoni   翻译：王玲

巴瑞洛切市是阿根廷巴塔哥尼亚省最美丽的小城之一。这里山水相依，植被丰富，景色旖旎。纳韦尔瓦皮湖占地面积超过 500 平方千米，一汪湖水碧如蓝，是当地居民和游客心驰神往、流连忘返的自然殿堂。

圣彼得半岛是整块陆地伸入湖中的部分，其悬崖峭壁提供了极佳的观景视野。正是在半岛上一处特许的地块上，建筑师 Alejandro Bustillo 于 20 世纪 40 年代修建了自己的房子和工作室。Alejandro Bustillo 是阿根廷最负盛名和最具创作灵感的体现现代动向的建筑师之一，除了在布宜诺斯艾利斯的许多设计之外，他最有名的作品是巴瑞洛切市融合了当地传统元素的 Llao-Llao 酒店，他还将这种元素运用到自己房子的设计之中。历经几十年的岁月变迁，Bustillo 的房子由买房人的孙子继承下来。在过去 5 年间经过一系列改造和修葺，再现了房子最初设计时的风采。

现在的业主决定在尊重巴塔哥尼亚省典型本土建筑风格的情况下重新扩建这座房子，营造出更多的花园空间，并保留乡村自然景观的质朴与古拙。为了实现这一设计目标，他邀请了布宜诺斯艾利斯的景观设计师 Martina Barzi 和 Josefina Casares，因为他们的设计风格与他本人及原来业主对当地景观的理解和热爱不谋而合。

根据设计师的设计，该项目需要数年才能完全

总平面图

建成，这个不断变化的过程要求在设计之初就做好决策，而且随着时间的流逝更需要业主对该项目的持续关注。在这片幽雅清静、景色秀美的环境中，对生命和大自然伟大力量的尊重才是设计灵感的源泉。

当 Barzi 和 Casares 在 2002 年第一次考察项目基地时，就认为整体环境不协调，甚至有点杂乱无章：业主们根据自己不同的喜好种植了不同的植被；大量的非本土松树阻挡了欣赏湖面的视野；入口道路生硬地插入项目腹地，影响了房前草坪的延展和对地块的清晰分区。因此，设计师提出了两项任务：首先，重新规划交通动线；其次，对场地进行全面清理。清理工作耗时近 1 年，仅清理浓密的花旗松就耗费了相当长的时间。花旗松是一种外来的松树，它们侵占了一部分本土植被的生存空间。设计师采用假山毛榉等本土植被代替那些被移走的树木。设计师在研究地块的原始特征和建筑师 Bustillo 对地块的理解时，发现了最重要的一点，那就是建筑师对场地的气候条件和土地状况的热爱与尊重，例如 Bustillo 在某种程度上对"风"——巴塔哥尼亚省标志性的景观元素予以了保留，在嶙峋的峭壁和绵延的海岸上，人们都能聆听到海风的耳语。随着时间的推移，松树在不断成长，使得拂面的清风与歌声或多或少地失去了原有的韵味。正是为了还原纯正的"风"这一景观感受并扩大观景视野，设计师花费数月将原来的植被移走，重新种植了许多适宜的新植被。

同期开展的另一项重要工作是将原来的道路重新规划，不仅使游客可以行至周边的树林，而且他们可以感受到场地的起伏，营造出步移景异的自然交通动线。在场地的高处，设计师充分利用开阔的视野，并结合当地的景观小品，让人们在那里欣赏美景、放松心情。树干制成的桌椅或聚集一处，或分散开来，为这里的极致美景平添几分古朴的气息。

设计师缩短了原来的入口道路，保留了花园和草坪用地，并在房前开辟出了大片空间，这片空间设计

137

得丰富多样、富于趣味。首先，环形空间的中心种植一些松树，房屋掩映在绿树丛林之中，若隐若现，充满神秘感；其次，树林的右后方是两块大岩石，标示出房子的所在，人们在此拾级而上便可到达房屋的主立面。项目施工之初，这两块大岩石上覆盖着泥土和杂草，遮挡了主立面的一部分视野。为了扩展视野以及更好地展现房屋结构，设计师决定清除覆盖在岩石上的泥土和杂草，并在建筑和自然之间形成一定的层次和平衡；房子的后面高高耸立着该场地上最为壮观的景观元素——一块近20米高的巨型岩石，展现出大自然的鬼斧神工。

这块巨大的岩石与一处人工建筑平行排列，巧妙地实现了自然与建筑的完美对话。建筑师Bustillo精心设计建造的工作室是一处布局紧凑的、中世纪风格的石质塔楼，仿佛是悬崖边上的一尊雕塑，淡泊谦和。环绕两块岩石和线形台阶种植着一些本土的野生花草，

它们与远处的塔楼相得益彰。环绕岩石种植的植被都是一些巴塔哥尼亚草原上的典型植物或草种，主要是Nasella tenuísima、神圣亚麻属和景天属植被。其他植被还包括本土观赏物种、移植的植被和其他一些同类的植被。

设计师在房子附近营造出一种草场景象：大片的草坪绵延起伏，房子两边种植着麦氏草、野生雏菊、羽扇豆等；各种紫色调的"地毯"与翠绿的树木、碧蓝的湖水交相辉映；金黄色的小草随风摇曳，呼应了最初对"风"进行应用的设计理念。越是靠近树林，植被就愈加显得原生态，地面点缀着繁星般的橙色和黄色的六出花以及紫色的毛地黄。

由于这里处于半岛的最尖端，欣赏湖面美景的视野开阔。在靠近悬崖峭壁的地方，设计师修复完善了现存的一些由当地石材建造的低矮挡土墙。建筑正下方的峭壁自然起伏延展，在寒冷的日子里显得清冷荒

凉、毫无生机，但是春末夏初、万物复苏之时，这里又绿意盎然。这里主要的植被是金雀花，这是一种发散状的灌木，金黄色的小花团团簇拥，甚是美丽。

与这种野生景观不同的是，建筑周围留有更多人工修饰的痕迹。起伏的地势在房子和花园之间突然改变，而且新的规划方案提出沿着房屋两个较长的立面设置一些露台或平台。这些平台同该场地的其他地方一样采用石质边缘，形成统一的建筑风格；它们不仅充当了建筑的"地基"，而且成为天然的墩座。在主立面一侧，这个隆起的平台展现出整个场地中惟一的几何种植模式。一间建有露台和工作区的温室以及设有木桌椅的烧烤区，为整个项目画上圆满的句号。

自然与历史成为设计的主旋律，景观也以一种微妙、恭敬，甚至出其不意的方式与其紧密相连。巴塔哥尼亚省当地典型的色彩、纹理、造型和表面以及才华横溢的建筑先驱的品味，都在此展现得淋漓尽致。

Bariloche is one of the most beautiful towns of Argentinean Patagonia, where lakes, mountains and native woods wonderfully combine to craft a place of stunning nature. Nahuel Huapi Lake, with more than 500 km^2 of deep blue waters, becomes one of the most impressive natural references for locals and visitors.

Peninsula San Pedro (or Saint Peter Peninsula) is one of the land formations which jut out into the water and whose steep stony edges provide the best vistas toward the rest of the area. It was right in the most privileged lot of this peninsula, that architect Alejandro Bustillo built his own house and studio, in the 1940s. Alejandro Bustillo was one of the most recognized and sensitive Argentine architects who represented the Modern Movement, and the one who designed, besides other many important works in Buenos Aires city, the well-known Llao-Llao Hotel in Bariloche. The vernacular image with which he designed this building is the same one he used to lay out his own place; inherited, after decades, by one of the grandsons of the person who bought the property to Bustillo, this place has undergone a process of renovation and revitalization, during the last five years, with the same sensitivity that had characterized the original design.

The present owner decided to enlarge the house, respecting the vernacular architecture of the house, so typical of Patagonia, and also to give the extensive park and gardens that accompany the construction the chance to flourish again, preserving the sense of a rustic natural landscape. With this objective, he called landscape designers Martina Barzi and Josefina Casares, based in Buenos Aires, whose work possesses the same kind of understanding and love for local landscapes he shares with the original owner.

This project is, according to the designers, one that takes years to establish completely, an ongoing process which depends not only on the first design decisions but on the careful attention that the owners pay to it as time passes by. In such an overwhelming, yet peaceful, natural enclosing, this appreciation of the greatest forces of life and nature are the ones which work out best.

When Barzi and Casares first arrived to the site, in 2002, its general image was inconsistent and even confusing: different plants have been incorporated by different owners and in different moments with no criteria, masses of non native pines blocked out the views to the lake and an access road extended far into the lot denying the

possibility of meadow-like extensions in front of the house and fractioning the land incongruently. Consequently, the two actions to be initially carried out were, first, the new outline of the access road and, second, the general cleaning of the site. The last took almost a year, a long phase during which the first works consisted in clearing the site of dense groups of Pseudotsuga menziesii or Douglas-fir, a kind of introduced pine which had invaded part of the native woods; however, the clearing process was slow and methodic, replacing some of the removed trees with Nothofagus dombeyi or Southern beech, or other native ones. One of the things that the designers had found most relevant, while doing research about the original character of the site and the ideas of architect Bustillo for it, was his love and respect

for the inherent climatic and land conditions of the place; the wind, a local mark of all Patagonian landscapes, was something that the original designer had wanted to preserve and be able to experience from the rocky cliffs and shores of the peninsula. With time and the proliferation of pines, the experience and sounds of the wind blowing in the lot had missed some of its original attractiveness and primitive sense; with the goal of providing this experience again and also of opening up some views to the water, they worked for many months removing old and planting new tree species.

Other significant actions, implemented during this same period, were the tracking down of old paths and the tracing of some new ones that would allow visitors to get into the surrounding woods and to go up and down the natural

slopes of the site, creating in some cases a surprising natural circuit. In some of the highest points, they took advantage of the exhilarating views and incorporated vernacular furniture, for people to just sit around, watch and relax. Tables and chairs, groups of benches or isolated ones, carved from tree trunks, add a quite pristine image to the already faultless setting.

The length of the access road was reduced, preserving the land for gardens and florid meadows and also leaving a significant larger space in front of the house. This space was designed to provide a diverse and interesting entering experience to the site: first, with a new roundabout whose central area is planted with some of the remaining pines which conceal the house and create an unanticipated sense

1

of arrival and, then, right after this woody area, with two large pieces of rock that finally mark the presence of the house and frame a plane of steps that lead toward its main façade. These two pieces of rock were covered with earth and grass when the project began, partly obstructing the view of this façade; in order to open it up and generate a framing structure for the house, the designers decided to clean it and, in this manner, established certain hierarchy and balance between construction and nature. Behind the house, and as the most imposing element of the site, an existing almost 20 meter-tall voluminous rock stands as a counterpoint to this arrangement of natural rocks.

However, this huge rock is paralleled with a manmade construction that finely completes the dialogue between architecture and nature; specially designed and built by architect Bustillo in the 1940s, as his own studio, a compact medieval-looking tower made of stone stands as a landmark right on the edge of the cliff and appears, stoically, on one side of the house. The planting that surrounds the two rocks and the linear steps, composed of native grasses and some wild native flowers, helps to define and outline this tower in the distance. The species chosen to encircle the rocks relate to the emblematic plants and grasses of the Patagonian steppe: Nasella tenuísima, Santolina and Sedum sp are the most important.

The rest of the planting plan relies on the use of native ornamental species or some naturalized ones and a pattern of large homogeneous masses of them. For the areas closer to the house, the designers chose a meadow-like image: some large surfaces of lawn and, at the sides of the construction, borders of Molinia, wild daisies and Lupinus polyphyllus or garden Lupin. Carpets of varied hues of purples contrast with the bright green of the woods and the distant incredible blue of the lake; in some other areas, the golden grasses sway to the breeze and honor that original idea of the presence of the wind. When getting closer to the woods, the planting plan gets somehow wilder and the ground is dotted with the orange and yellowish flowers of the Alstroemeria aurea or Peruvian lily and the fascinating purple of the Digitalis purpurea or Purple foxglove.

Because it's the furthermost tip of land of the peninsula, this lot faces the lake around almost the complete extension of its perimeter; on those areas adjacent to the cliffy edges, the designers restored and completed some existing short retaining walls, made of local stone. Right below, the natural slopes of the rocky cliffs drop away; during cold months they look bare, but right before the beginning of the summer these surfaces become profusely conquered by the native flora, mostly composed of Genistas or Dyer's greenweed, a spreading bush which blooms spectacularly with golden-yellow tiny flowers that reproduce in huge masses.

In contrast with this wilder image, the areas which surround the house become more formal and manicured. The existing slopping of the site produced some abrupt changes between the level of the construction and that of the garden itself and the new plan proposed the creation of some "terraces" or flat planes which extend along the two longer façades of the house. These planes, which are architecturally defined with stony edges in the same manner and with the same stone used in the rest of the site, establish the "founding" of the house and mark a natural podium; in the main façade, this raised plane displays what turns into the only geometrical planting pattern of the entire site.

The plan is completed with the incorporation of a green house, with terraces and working areas, and a place for barbeques with wooden tables and benches.

In a site where both nature and history arises as absolutely significant and attractive components, the landscape proposal offers a bond with them, in subtle, respectful and sometimes unexpected manners. The colors, textures, shapes and surfaces of Patagonia are exhibited here, together with the architectural taste of one of his most sensitive pioneers.

绝壁上的田园风情 —— 智利Huentelauquen住宅

Bucolic Landscape on the Cliffs — Huentelauquen House, Chile

撰文：Jimena Martignoni　　图片提供：Teresa Moller Office　　翻译：申为军

Teresa Moller 是一名屡获国际大奖的智利景观设计师，在她的每一个项目中，景观与自然的和谐似乎都成为设计的主导因素。当涉及到住宅项目时，对景观的理解和体验则上升为对居住者的一种引导方式，即帮助他们理解和接受如何以更具价值、更协调、更和谐的方式利用、展示并优化原有的景观，而非占有和强行改变。

Huentelauquen 住宅项目就是为住宅创建辅助景观的绝好范例。该项目位于智利首都圣地亚哥以北400km处，一幢房屋孤零零地构筑在乱石嶙峋的海滩上，被完全隐藏在面朝大海的悬崖绝壁上。在点缀着仙人掌科植物的沙地草原上，一条私人公路通向这里，沿途几乎看不到房屋。只有在靠近或抵达海滩后，才能看见那幢似乎悬挂在绝壁上的房屋。

Teresa Moller 在房屋建造的初期便参与其中，以上情形就是她在设计时所要考虑的因素。Teresa Moller 最早做出的几个决定之一就是要利用住宅屋顶——因为屋顶与悬崖的高度几乎一致，可以视做景观的自然延伸，或者作为从视觉和整体环境上保护现有景观的一种方法。此外，她还决定充分利用该地区现存的本土植物，以此加强住宅与该地区特有景观要素的联系。这块比较孤立的景观主要为草原，只有少量的植物品种。在该项目中，智利所特有的植物品种（Baccharis linearis, Baccharis concave, Puya chilensis, Calandrinas）被集中种植，并呈现出自然生长和扩散的态势，形成大面积的绿褐色，与远处的海景形成鲜明对比。对于看惯修剪整齐的花园景观的人而言，这样的种植方式显得有些凌乱，因为植物的长势有所不同，有些矮小短粗，有些则浓密繁茂。但不管怎样，这些地被植物都与周围的海洋有着显著的区别。

从屋顶到下面的沙滩有两条小径。设计师试图使居住者在下山的途中充分体验到现有景观的壮美，设计精巧、宛若天然的下山小径看上去似乎一直就存在于这里。

其中一条小径由一连串的石阶组成，依地形和坡度而下，很是引人注目。白色的石块被精心垒砌成一个个台阶，硬朗的几何线条与周围景观的有机形式以及大片仙人掌科植物形成强烈对比。这些数量众多的仙人掌科植物具有建筑和雕塑般的独特之美，屹立在石阶两侧，成为另一种效果显著的设计要素，保持了场地的活跃性。石阶和仙人掌植物一直延伸到崖底，从那里开始逐渐消失，它的尽头即是沙滩和大海。

第二条小径不是很陡峭，由一组不规则的踏步或平台构成，使下山的路更为平缓易行。路边的植物配置与屋顶相同，只是减少了仙人掌的数量，同时在面向大海的斜坡上增加了水平方向生长的低矮植物。

该项目最重要的一个方面是打磨、展示、保存现场存在的天然巨石。房屋的一面构筑在海边的一块陡峭而巨大的岩石上。在建造初期，这块露出地面的岩石几乎全部被泥土所覆盖。为了展示其原有风貌，花费了整整一个月的时间对其进行清洗、打磨。在这一过程中，岩石的表面逐渐显示出赭色、白色、棕色、金色等丰富的色彩，并从物理和视觉上都表现为房屋的支撑点。岩石在房屋前形成一个大型的天然露台，这使得设计师能够运用一定的设计元素，在景色最美的地方布置那些可供漫步、眺望、休憩的设施。

通过简单的几何式布局，设计师确定了一种以岩石为支撑的步道—码头的形式。因为邻近大海，这个构筑物首先是座天然码头，守望并俯视着大海。同时，它也是一个可供通行的步道，并起到指引的作用，提供了独特的视角。它由很多原木构成，排列并固定在岩石表面，由此形成中央空间并被设计成火坑，冬季，人们可以围坐在周围取暖，以抵御海边的寒冷和潮湿。此外，设计师还在各个角度安排了几组原木长凳，方便居住者小坐和眺望远方。

岩石不断地出现，但并非仅仅起到定位的作用。在房屋建造期间，住宅内部的地表显露出另一块重要的岩石，它同样被清洗和保留了下来。这块岩石占据了中庭一半以上的空间，与住宅的入口斜坡相连，周

围铺设了木质平台并点缀了少许小型的当地植物，形成一个完整的构图。木质平台的边缘并不与岩石相接，而是在岩石四周留出一定的空间，仿佛守护出土文物般小心翼翼。

通往住宅的公路有几千米长，沿着海岸穿过草原。路边分布的石块说明已经抵达目的地——靠近住宅的若干组 6 块 ~ 7 块尖尖的石块就是入口标志。就像其他的一些标志性景观元素（如史前竖石、石柱等）一样，这些石块巧妙地表达出公路、抵达和边界的概念。

总的来说，该项目的景观规划具有如下特征：设计巧妙精湛、造型、材质和构图朴素简洁。这些设计要素构建出一幅自然的田园景观，但它又极富意味，与建筑本身的几何形式形成对话和交流，使整块场地获得绝妙的平衡感。而且，这样的场地也毫无疑问地证明了人类与自然的关系既冲突又密不可分。

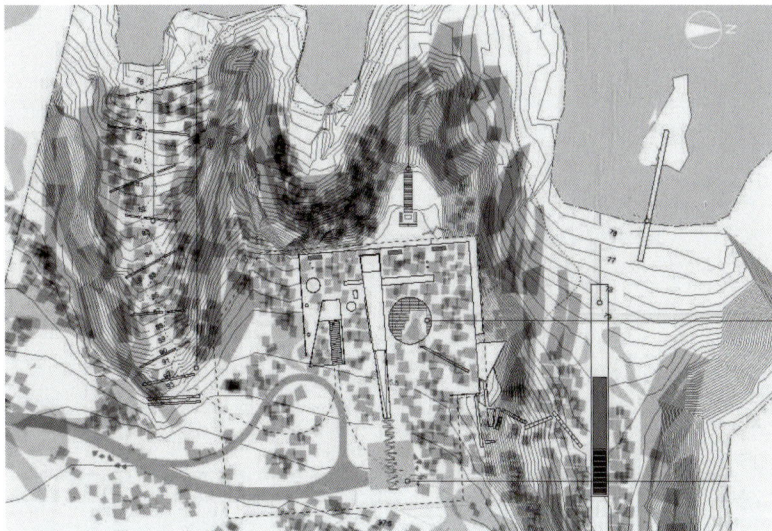

总平面图

In every project of Teresa Moller, a Chilean landscape architect internationally awarded, landscape and nature seem to be the guiding elements for the design. When it comes to residential projects, the experience and comprehension of landscape arises as a manner to educate the dwellers of the house, giving them the tools and motivation to understand and accept that is more valuable, more coherent and more balanced to lean on the landscape, framing it and enhancing it, than any other way of appropriation or modification.

The project of Huentelauquen, 400 km North of Santiago de Chile, is about the creation of a supporting landscape for a house which, isolated in a rocky beach, is completely buried into the natural slopes of the cliffs that face the ocean. When getting to this place, going through a private road that develops along a sandy steppe dotted by cacti species, the house is not visible. The visual contact is only possible when getting closer and going down to the beach, where the construction appears as if hanging from the rock.

This is the situation that Teresa Moller took into account to model the project, to which she joined when the house was under the first phases of construction. One of the first decisions she made was to use the roof of the house, which coincided with the ground level of the cliffs, as a natural extension of the landscape; or, actually, as a way to preserve it, visually and environmentally. However, she also decided to incorporate the presence of the native plants of the area, thus establishing a stronger link with the specific elements of this landscape. Mostly recognized by extensive steppe-related areas, the enclosing landscape presents only a few species which dot the ground; in this project, these species are grouped and create a green-brownish plane that contrasts with the distant image of the sea: Baccharis linearis or Chilean romerillos, Baccharis concava, Puya chilensis or Chagual and Calandrinas are the main species planted here, which grow and spread out naturally. For those who relate a garden with a manicured landscape and expect to meet with this kind of image, the view of this green plane could be perceived almost as an untidy one, where plants grow in diverse sizes, sometimes stumpy and sometimes voluminous; in any case, the ground cover is dramatically distinguished from the maritime background.

From this plane, there are two options to get to the sandy planes of the beach, below. Teresa Moller sought to offer the possibility of getting there through a sensitive experience of the existing landscape, a subtle and spontaneous descending path which seems to have always belonged to this place.

One of the descending paths reveals noticeably on the landscape, with a flowing series of steps made of stone which adjust to the topography and natural shapes of the slope. The white stone is nicely piled to model every one of the steps and create geometrical lines that contrast with the organic ones of the landscape and the homogeneous groups of cacti. These numerous cacti – architectural, sculptural and picturesque – multiply along the edges of the stairs, accompanying and framing it, as another evident design component and, also, keeping alive the existing image of the site. An image of unfolding stone and vegetal forms,

naturally reproducing on their way to a last plane at the bottom, where they start to vanish and give place to the sand and the water.

The second descending path, which is considerably less steep, is built as a series of irregular steps or planes that facilitate the way down in a more subtle and, probably, more submissive manner. Here, the planting repeats that of the green roof, with some less cacti and more horizontal-looking shapes displayed onto the sea-facing slopes.

One of the most important tasks of this project was the polishing, exhibition and preservation of the existing rock. One of the only two façades of the house leans on a plane of rock which drops steeply away toward the beach; at the moment the construction was beginning, this outcropping rock was almost completely covered with earth and, in order to exhibit its original surfaces, it was cleaned and then polished during approximately one month. After this cleaning process, the rock displayed an intense variety of ochre, white, brownish and golden hues and now appears as the physical and visual support of the house; extending in front of it, before dropping away toward the beach, the rock becomes a large natural terrace. For this reason, the designer incorporated certain elements that offer and make easier the possibility of sauntering, watching and relaxing in a place of particular beauty.

With a simple geometrical layout, Teresa Moller defined a path-pier that rests on the rock. This piece is, indeed, a pier, because it gets closer to the sea, looks for it and persecutes it, even when it's a passive persecution and it only stays, immutable, on the rock; and this piece is, indeed, a path, because allows people passing, guiding and bearing them, even when it's a short walk which becomes also a privileged view-point. Built with logs, aligned and fixed to the rocky floor, this path-pier also frames a central space which configures a fire pit, around which is possible to sit and get warm during the typical cold and humid days of the local maritime winter. On the side, a group of isolated logs, disposed as benches in different angles, invite to sit around and look into the distance.

Yet, the rock is not only a bearing element but it appears and reappears constantly. When in the construction process, another important piece of rock outcropped inside the house's surface. The decision was, once again, its cleaning and preservation. This piece occupies more than half of the underground central yard, connected to the accessing ramp of the house, and is complemented with a wooden deck and a few small native plants which timidly complete the composition; the edge of the deck never really touches the rock, it retreats and encloses it, as if indicating an archeological finding. In fact, revealing it as one.

The road that goes to the house, running through the steppe after having gone for kilometers along the ocean shores, displays a series of rocks that mark the arrival: when getting close to the house, some groups of six or seven pointed rocks appear as accessing signs, or as milestones. With a fine reference to the emblematic elements that mark the action of displacing into the landscape–menhir, dancing stones, monoliths, moabis—these rocks implement, subtle but eloquently, the concept of road, arrival and boundary.

The landscape plan for the Huentelauquen House is characterized by a sense of subtleness and sensitivity, sober forms, materials and compositions; this combination, realized in a highly bucolic landscape, yet meaningful, and in close dialogue with the strong geometry of the architecture, results in a place with an especial balance. What's more, a place which clearly proves how controversial and passionate is the relationship between man and nature.

法夸尔住宅

Farquhar Residence

撰文：Lisa Jenkins　　图片提供：Mark McWilliams　Tom Jenkins　　翻译：牟誉

田园式的草坪、精致的布景以及大量的鸡爪槭勾勒出了庭院的整体景致，体现出设计师将得克萨斯风情与日本风情完美地融合在一起的设计理念。新建的住宅、水池、会客厅以及原有的游泳池共占地8093 ㎡，对面是一个公共池塘。漂亮的房屋和池塘、开阔的草坪以及保留下来的树木都属于私人财产。新建住宅坐落于两株茂盛的榆树之间。庭院的主人爱好收集盆栽植物，但他不喜欢规划整齐的庭院道路，而喜欢在宽阔的草坪上漫步，欣赏开阔的庭院景色。

该住宅的特别之处在于设置一间用来存放和维护盆栽植物的盆栽坊，并收藏了大量历史悠久、有价值的盆栽植物，其中包括一个具有400年历史的稀有植物品种。设计师花费了3年时间来寻找能够抵挡得克萨斯州多变气候的羽毛枫和其他特别的鸡爪槭品种，并将它们分布在庭院的各个角落，像安装户外雕塑一样。设计师将单株植物种植在石头底座中，将植物成组地放置在两个设计独特的长凳上——一个长凳用做植物维护，另一个长凳则用做植物展示。长凳放置的位置符合园艺学的要求，充分考虑了光照和周围的小气候。精致的细节设计以及干净简单的布置突出了院落空间的特点，大量栽种的鸡爪槭也丰富了主人的植物收藏。设计师还重新安放了原有的 Jesús Moroles 雕塑，为庭院和主房屋的景致增色。

庭院主人对泊位附近石板路的改造非常满意，原来泊位附近杂草丛生，现在则种植了蕨类、小盼草和黄菖蒲，在水边形成了一处新的风景区；住宅周边的种植区为主人的隐私提供了保障。

为了与庭院的自然环境相协调，设计师用Ozarka 石头和彩色混凝土铺设了独具匠心的行车道，石板之间还种植了沿阶草来进一步软化路面。主房屋和客房之间的行车道穿过草坪通往池塘，使视野不仅仅局限于房屋前的范围。

总平面图

The landscape architect's concept—Texas meets Japan—manifested itself in the pastoral lawns, intricate details, extensive selection of specimen Japanese Maples, and controlled views throughout the grounds of the client, a bonsai collector. The new home, pool and guesthouse, and existing swimming pool sit within a two-acre double lot facing a pond, shared with a dozen other homes. This collective of houses and pond is park-like and the client's property included expansive lawns and trees protected during construction. The new home is positioned between two mature American Elms. Unlike many clients who want orderly and defined garden paths, this homeowner enjoyed strolling across the open lawn and preferred unobstructed views of a green.

Unique to this residence is the client's bonsai house used for the potting and maintenance of a large collection of old and valuable bonsai plants including a rare 400 year old species. The landscape architect shopped for three years looking for exotic Japanese Maple species of Dissectum and other varieties that would tolerate the erratic North Texas climate. The landscape architect displayed the plants throughout the grounds like an outdoor sculpture installation

with single plants set on stone vitrines and groupings arranged on two very specific bench designs, one for plant maintenance, and the other for display. The bench placement within the garden was horticulturally sound based on light and microclimate concerns. Subtle details and clean, simple lines characterize the space while concentrated masses of Japanese Maples add to the client's plant collection. The landscape architect reset an existing Jesús Moroles sculpture to increase viewing options from the grounds and main house.

The client did agree to the reconstruction of a flagstone path near an old, overgrown boat launch now planted with ferns, Inland Sea Oats, and Water Iris creating a new refuge near the water. Formal planting beds are kept to a minimum and concentrated around the perimeter of the lot to address the client's want of screening and privacy.

Like the subtle nature of the grounds, the landscape architect crafted a highly detailed driveway using bands of Ozarka stone and colored concrete further softening the driveway surface with Dwarf Mondo Grass planted between the joints. The driveway placed between the main house and guest quarter's offers open and controlled lines of site across the lawn toward the pond rather than emphasizing views of the front of the house.

1 纵横交错的道路
2 石板小路
3 重蚁木的楼梯和地板
4 温泉房的入口
5 游泳池
6 行车道
7 盆景房
8 Jesús Moroles 雕塑
9 银杏盆景展示台

城市生活对山、水栖居的追求 —— 深圳龙岸居住区

The Urban Pursuit of Residence by Mountains and Water — Joyful Palace, Shenzhen

撰文：深圳市华域新实践国际景观设计有限公司　　图片提供：张阳　危薇

1　滨水栈道
2　水岸夜景

该项目在 2009 年~ 2010 年市场推出的高档楼盘中引起人们的广泛关注。设计师将体现出该项目的独特价值及不可复制的气质作为核心理念。

上风上水中轴线

该项目位于深圳南北中轴线向北延长段，南面是水库保护区，基地呈两峡一谷的形态，整个场地高于周边地块。该项目根据上风上水的风水格局潜心设计，精心建设打造，将高层区沿外围依两山布局，多层和高档别墅区在中心区，景观设计因势利导建造景观湖形成核心景观区，并以湖、山为基本元素，淋漓尽致地诠释了依湖而栖、傍山而居的生活特质。

含蓄向心的景观格局

山湖格局及高层在外、多层在内的基本布局，确定了景观序列的含蓄向心的特质。

由项目主入口进入，穿过浓阴遮蔽的入口花园，伴随着气势磅礴、潺潺而下的跌水，到达台阶的上方后，成组的加拿利海枣迎面而来。视线向前延伸，湖已在近前，波光涟涟，花木葳蕤。城市的喧嚣聒噪早已不在，古榕下，滨湖的栈道旁，生活就在此处。

依湖而栖

别墅区沿湖展开，并向南北两侧坡地延伸。中心湖景为该项目的核心景观区域。此设计着重于展示湖的景观价值并强化湖与人的活动的交流。设计师以会所为背景设置了亲水广场，由亲水广场延伸出沿湖的公共木栈道和上山的主通道。沿湖北岸和西岸的别墅均设有私家亲水平台，使人们能够与水亲切对话。设计巧妙地利用绿岛和跌水造景将水景延引到会所地下

室的顶板上，在技术上解决了水景"跨缝"的难题，形成一个充满动感的浅水区，也使人们走出会所便可欣赏到水景。

傍山而居

基于该项目两山夹谷的格局，景观设计完成了南北两处"山·坡地"的景观构想。北侧区域由湖滨亲水平台傍山就势向上延伸。步行区在不同标高的区域间流畅通达又开合有致，且与车行道路完全分离。在步行区重要转折点设有休憩平台，并利用其标高优势组织视线，使人们在行进中步移景异，并可眺望远景。

在两山格局的南山，设计师利用更大的空间尺度和自然原土区域，创造了一个林木环抱的溪谷景观区。从谷地望向高处，林木茂盛，潺潺的溪水由远而来，人行散步道蜿蜒其中。谷地的顶端设有一眺望台，临

手绘局部平面图

谷而立，清风徐徐。台侧有瀑布，水涓涓涌出，之后沿山谷曲折而下，从花丛林下穿出。植物配置、场地处理与建筑相互融合且细致入微。

在台地花园中，设计师以溪水将各处不同标高的花园连成一体，并在各层之间设置高档提升电梯，体现人性化设计。

由该项目南望，瑭郎山一线及水库区景观宛若在咫尺之间；向该项目的中心看去，中心湖景，溪谷山景尽收眼底。该项目的景观设计旨在体现一种能表达当下生活特征、体现项目独特性格的创造性途径，并将之贯穿于该项目的整个过程之中。

Shenzhen Long'An Garden, one of the top rated building dishes released in 2009-2010, stirs highly attention in the market. Its unique value and unduplicated features serve as the main design concept in its landscape.

Central Axis of Perfect Feng-shui

Located right in the northern extension section of South-North central axis, Long'an Garden faces the reservoir protection area in the south. The whole site is formed of two gorges and a valley, much higher than ambient plots. Based on this perfect Feng-shui structure, high-rises are elaborately laid along the two gorges, multi-storey and first class villas placed in the center. Geographically, the central view area perfectly interprets living features through basic natural elements of lake and hill.

Landscape layout of Convergent Centralization

The basic layout (of lake and hill, high-rise outside and multi-layer inside) defines the landscape feature of convergent centralization.

Enter the site from main entrance; firstly pass through the entrance garden of thick shadow, grandeur cascade trickled down, and then reach upper steps, groups of Phoenix canariensis assembled to welcome guests. With your view-

1　中心湖全景
2　入口雕塑及水景
3~6　局部小景 1

sight forward, the lake is right in your front, ripples bright and flowers blooming. Urban noise faded till disappeared, only left the life in ancient Banyan and lakeshore board walk.

Resting by Lake

Starting from the lake, the villa area stretched itself along the southern and northern sloping fields. As the kernel landscape area, central lake view is laid out with functional apparatus aiming to fully demonstrate its landscape value and help strengthen communication between lake and people here. Public board walk and main up-mountain passage extend from the waterfront Plaza set in the clubhouse. Waterfront terrace of north and west lake shore villas offers people the platform to close to water. Layout of Greenland and Cascade skillfully draw your attention down towards the crest slab of clubhouse basement. It not only resolves the problem of "water gap" but also forms a dynamic shallow water area which demonstrates water features in front of visitors upon coming out of the clubhouse.

Resident by Mountain

The landscape design realize the conceive of hill and slope view in the south and the north. In northern area, the waterfront terrace by lakeshore stretches up along the hill. Pedestrian road is completely separated from driveway and makes smooth transition between different level areas. Resting platform set at the critical transition of walking steps, making use of different level advantages, people may enjoy near view in leisure walking and look out of a distant view as well.

In the south, the designer utilizes larger space dimension and natural geographical advantages to create valley scenic area in trees. Seen from the valley, lush trees and trickled stream wind from a distance to nearer with pedestrian steps wandering in it. Top of the valley is a viewing tower standing at the entrance and welcoming gradual freeze. A waterfall trickles from one side of the viewing tower, windingly down along the valley and finally run out though flower bush. Plants furnishing, geography treatment together with its buildings fuses with each other in tiny subtlety.

Stream in Terrace Garden integrate different level parts into a whole. Elevator is set on each floor to make convenience for people.

Seen from the north site, Tang-Lang Mountain and reservoir scenery is near almost within reach. Here you may have a panoramic view of the central lake and valley. Design theme of Long'an Garden is to demonstrate modern lifestyle and creativity upon unique project characteristics, and runs them in the whole project realization.

旧金山自由山住宅

Liberty Hill Residence in San Francisco

撰文：Deborah Bishop   图片提供：Marion Brenner   翻译：王玲

1 花园与远处的城市遥相呼应

1

花园平面图

旧金山自由山上的这座私家花园所处地形十分陡峭，设计师对硬质景观元素进行了创新性的处理，是一个集休闲娱乐于一体的成功花园的典范。考登钢制成的围护结构和种植床穿过周边的木栅栏沿场地周围一直延伸开来。Surfacedesign公司设计的后花园与起伏的城市地貌交相呼应，人们站在高处露台或从房子的后窗眺望，可将城市景致尽收眼底。

考登钢制成的种植床和混凝土墙被添加到原有的石挡土墙上，勾勒出一系列蜿蜒的Z字形小径，引领人们从屋内走向花园。建于上世纪中期的石墙不仅展现出了石材、混凝土、木材和钢材等自然的原真性，而且从视觉上将场地的过去与现在和谐地融合在一起。种植床有的高高在上，有的插入地面，有的则嵌入红木栅栏——阳光透过栅栏洒下斑驳的斜影，增强了种植床的几何质感。

为了充分体现花园的整体性，大片的儿童活动区都铺设了草坪，休闲露台也采用风化花岗岩。种植床不仅是主人施展园艺技术的舞台，同时也使园中蜂飞蝶舞、鸟语花香。混凝土台阶上的排水槽不仅增强了园内的排水功能，更突出体现了花园倾斜的自然地势，花园内的灯心草对流入地下水系统的雨水起到了一定的净化作用。由于种植床周围也设计了排水槽，这些种植床就仿佛从地下长出来一般，增强了自身的纵深感。

主人不仅喜欢置身园中，更喜欢在室内俯看花园，而观赏花园的这份闲情雅致又不时地被鸡爪槭或者蕨类、鸢尾属、荚蒾属和银莲花属等喜阴植物所打断。

花园中的植被：鸡爪槭、茶梅、山茱萸；种植床中的植被：铁线蕨、银莲花、穗乌毛蕨、风铃草、嚏根草、鸢尾、狗脊蕨；周边花园中的植被：绣球藤、木百合、日本绣球。

Marked by its dramatic sloping topography, this private garden in San Francisco's Liberty Hill has dedicated areas for entertaining and children's play, which are defined and navigated by an innovative arrangement of hardscape materials. Cor-Ten steel boxes serve as retaining structures and planters, extending along the site's perimeter and penetrating the surrounding wood fence. Surfacedesign Inc. approached the whole back yard as an abstracted allusion to the hilly city itself, which appears to fall away behind the garden when viewed from the upper terrace or out the back windows of the house.

The Cor-Ten boxes and concrete walls were added to existing stone retaining walls to create a series of switchbacks that guide the journey from the house into the gardens. Erected mid-century, the stone walls form a natural part of the vocabulary of rugged and refined materials—stone, concrete, wood, and steel—while uniting the site's past and present incarnations in a visually harmonious way. The steel boxes, which harbor plant life up top, descend into the ground plane and insert into the redwood fences—a geometric composition enhanced by the slanting shadows that are cast through the variegated slats.

Seeking to maximize permeability for the overall health of the garden, the large children's area is lawn and the entertainment terrace is decomposed granite. The planter boxes not only afford the homeowners an opportunity to do some hands-on gardening, they also attract pollinators such as birds and butterflies. The concrete steps are framed by runnels to enhance drainage (while highlighting the sloped nature of the garden), and a native sedge called Juncus helps purify run-off before it enters the groundwater system. Because the steel boxes are framed by the runnels, they appear almost to rise from below the ground plane, adding depth to the composition.

As the homeowners spend as much time looking down on the garden from their home as being in it, they also appreciate its sculptural quality, which is punctuated by plant material such as Japanese maples and the softening presence of shade-lovers such as ferns, irises, viburnum, and anemones.

Trees –

Acer palmatum 'Nishiki Gawa' – Japanese Maple
Camellia sasanqua 'Yuletide' – Yuletide Camellia
Cornus 'Eddie's White Wonder' – Eddie's White Wonder Dogwood
Steel Box Planting -

Adiantum capillus –veneris – Maidenhair Fern
Anemone hybrida – Japanese Anemone
Blechnum spicant –Deer Fern
Campanula isophylia – Bellflower
Helleborus niger – Black Hellebore
Iris douglasiana – Douglas Iris
Woodwardia fimbriata – Giant Chain Fern
Surrounding Garden Planting –

Clematis montana 'Rubens' – Clematis
Leucadendron 'Safari Sunset' – Safari Conebush
Viburnum plicatum Thunb – Japanese Snowball

彼得森住宅

Peterson Residence

撰文 / 图片提供：Marion Brenner James Lord   翻译：王玲

这座新的花园中蕴含着业主的双重期望——为孩子们提供广阔的休闲空间的同时展现出景观艺术的无限魅力。Surfacedesign 公司利用坡度平整技术、场地上的再生材料以及场地独特的元素创造出一座雕塑般的花园，它不仅是人们沉思、游玩、探索的理想场所，更融于当地的地域景观中。

Surfacedesign 公司设计的这座雕塑般的花园坐落在一个能够俯瞰旧金山湾的半岛上，显示了狂热的艺术收藏者独特的想像力。而这些收藏者的子女正在经历和享受的景观设计理念体现出他们的艺术创作激情。花园将几个独特的空间交织在一起，使建筑、景观和功能寓于其中，充分展现出场地的独特景致和地形特色。

场地入口处，隆起的木板路一直从房中延伸出来，使房屋与木兰、草坪和蕨类植物交融渗透在一起。木板路的翘面充当景观的起始符，强调出房屋原有地面和周围地形突变之间的联系。越过木板路，原来的鹅卵石小径与一个新种植的绿荫花园相连，为主入口处

营造出宁静私密的氛围。在入口主路的对面，一条主排水渠被当地河流中的石块巧妙地隐藏起来，并且石块与原来的斜坡浑然天成。穿过前门，葱郁的竹林和喜阴树木勾勒出一条蜿蜒的小路，小路的尽头是一个被草丛环绕的雕塑般的景观亮点。这种雕塑般的山坡景观沉稳大方，与大片的草丛相映成趣，形成一派如清风掠过般的当地景观。在通向东面草坪的道路上，可以眺望到旧金山湾和加利福尼亚州绵延山丘的迷人景致。重新使用的花岗岩边石材料将人们引至较低的草坪处，在这里绵延起伏的绿丘与远处伸入海湾的多山半岛相呼应。房屋的北面，常绿树篱和芬芳的多年生植物与园中的建筑交织在一起，营造出宁静的观赏花园；绿色树篱也是乔治·里基创造的动态雕塑旋转运动的一个抽象演绎。

儿童游戏区在整个场地中随处可见，场地入口处的两组隆起的木板路充当了孩子们的攀岩墙。儿童游戏区的亮点是设计巧妙的绵延绿丘，它们打造出奇妙的发现之旅和独特的游乐空间："兔子山洞"使孩子们可以将球投入到大小不同的管洞中，但却无法知道球会从哪个管洞滚出；"跳跳绿丘"是一张被隐藏起来的蹦床；"小矮人之家"位于游戏区外，是绿草如茵的理想私人游乐场所。同样，重新使用的花岗岩边石材料和弯曲的草坪筑堤也常常充当着孩子们户外即兴表演的剧场。

雕塑般的整体景观和业主对孩子们的喜爱之情与艺术收藏交织在一起，使得公园不仅是人们沉思、游玩、探索的理想场所，而且还淋漓尽致地展现出当地的景观特质。

This new garden weaves together the dual passion of the homeowners'—the desire for their children to have full range over the yard, while simultaneously displaying prominent pieces of landscape art. Through the use of grading techniques, reclaimed materials found on-site, and site-specific elements, Surfacedesign designed a garden that is a sculpture unto itself, providing unique spaces for reflection, play, and discovery, while tying into the native, regional landscape.

Situated on a peninsula overlooking San Francisco Bay, Surfacedesign designed this sculptural garden reflecting the unique vision of passionate art collectors whose personal artistic impulse includes the idea of the landscape being experienced and enjoyed by their children. As a result, the garden incorporates several distinct spaces that weave together architecture, landscape, and use, while seeking to capture the site's unique views and topography.

At the site's entrance, a folding deck project from the interior of the residence, merges the house with a landscape of magnolias, grasses, and ferns. The deck's warping planes create a threshold for the visitors, highlighting the relationship between the existing ground plane of the house and the sudden change of the surrounding topography. Beyond the deck, an existing cobbled pathway connects with a newly-planted shade garden, creating a sense of calm

and intimacy around the main entrance. Opposite the main entrance path, locally-sourced river stones artfully conceal a main drainage swale, while tying into the existing slope. Past the front door, lush groves of bamboo and shade-loving trees frame the lyrical movement of a meandering path that terminates in a sculptural focal point surrounded by hillside grasses. The solidity of this hillside sculpture juxtaposes with generous sweeps of grasses, referencing the native wind-blown landscape. The approach opens up to the east lawn area, offering extensive views of San Francisco Bay and the rolling California hills to the north. Re-used granite curbing leads the way to the lower lawn where a succession of green mounds abstracts the distant hilly peninsulas that penetrate the Bay beyond. To the north of the house, a sequence of evergreen hedges and fragrant perennials create a serene viewing garden, aligning with the interior architecture. The green hedging serves as an abstracted datum highlighting the twisting movements of a kinetic sculpture by George Rickey.

The folding deck at the site's entrance of Children's play areas doubles as a climbing wall for the active kids. The focal point of the children's area is defined by the playfully sculpted green mounds affording a wonderland of discovery and distinct play spaces: Rabbit Hole Hill allows children to drop balls into tubes of various sizes, without knowing where they will emerge in the landscape; Jump Mound conceals a hidden trampoline; The Hobbit House, situated beyond a play structure, gives the children a personal play space among a landscape of native grasses. Similarly, the re-used granite curbing and curved lawn embankment is frequently used as an impromptu children's theater for open-air performances.

The overall sculptural quality of the landscape weaves together the homeowners' passion for their children and art collection by providing places for reflection, play, and discovery, while acknowledging the native, regional landscape.

融入水元素的设计 —— 位于圣保罗的两处住宅项目

Incorporate Water as a Design Element — Two Residences in São Paulo

撰文：Jimena Mareignoni　图片提供：Isable Duprat Office Jimena　翻译：刘建明

1

黄色住宅总平面图

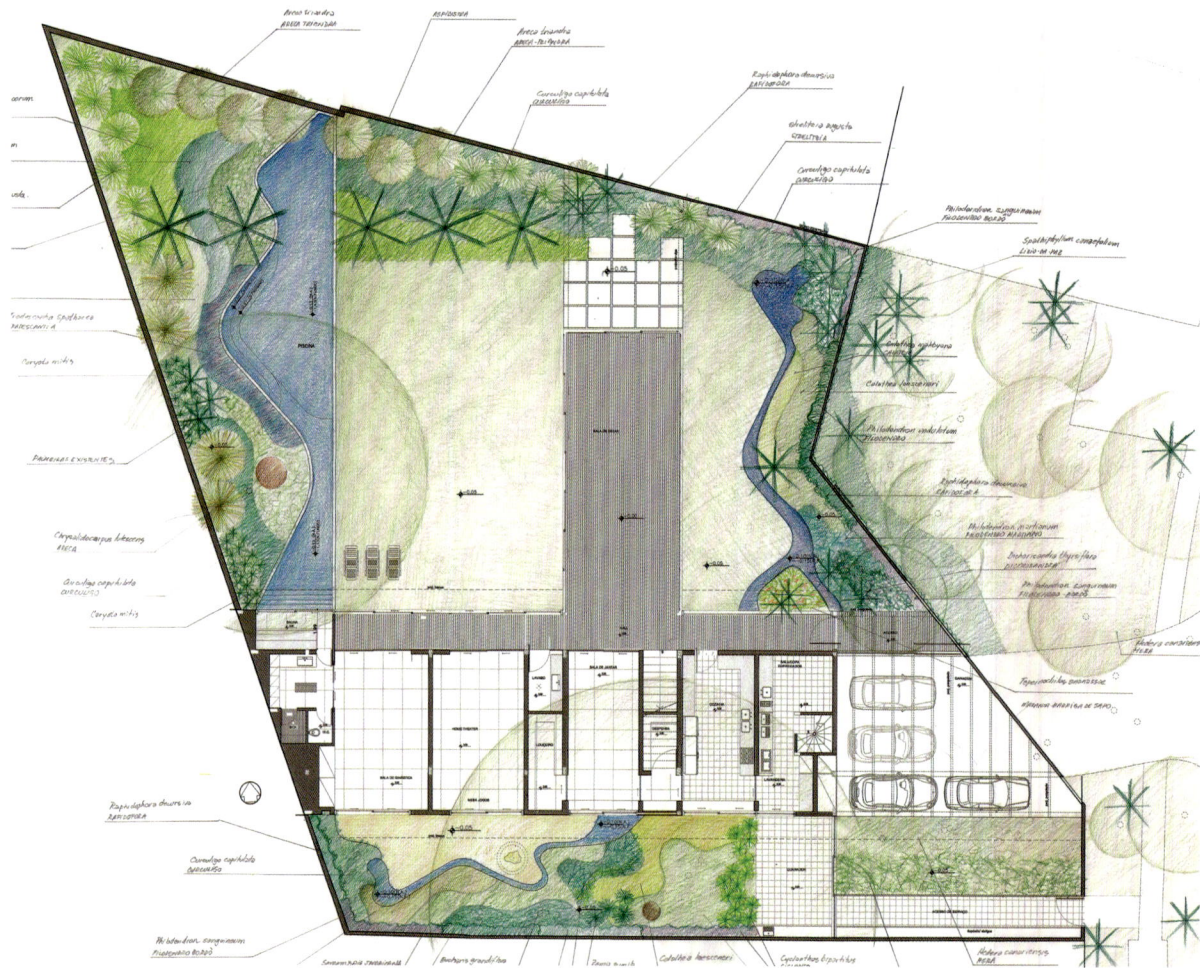

玻璃住宅总平面图

巴西圣保罗是南美最大的城市，同时也是世界第四大城市，其中 1800 多万人生活在城区，还有 1100 万人生活在市区（不包括郊区）。虽然圣保罗是以 1554 年创建这个城市的耶稣信徒的名字来命名，但是现在的圣保罗市民大多是来自不同国家的移民后裔。

从 19 世纪中叶开始，随着该地区周边咖啡种植业的逐步壮大，圣保罗的城区范围也急剧扩大，圣保罗首府的影响力也随之扩大，逐渐发展成为巴西的工业和金融中心。大多数成功的实业家都定居在一个名为"欧洲花园"（Jardim Europa）的社区，这里的街道布局整齐有序，豪宅与花园的设计都遵循极为经典的欧式风格。这些经典的豪华社区与圣保罗市内大多数的贫民窟形成了鲜明的对比，同时也是这个城市最不协调的一个特征。当然，和世界上其他地方一样，这些精雕细琢的高档社区也为城市增添了无限的魅力和风采，这里每一处私宅的景观设计都令人叹为观止。圣保罗的景观设计师伊莎贝尔·杜普拉特就是在此类项目的景观设计领域中颇有建树的一位大师，其中两处位于"欧洲花园"的私宅颇有创意——或突出私密性，抑或彰显开放性，这两处项目的最大共同点在于对水元素的巧妙利用。

黄色住宅

就在伊莎贝尔·杜普拉特 (Isabel Duprat) 应邀负责该项目的景观设计之际，业主又购得其附近的一块宗地，这给设计增添了难度。新购的地块要比原有场址高出约 1 米，因此，设计的最大挑战在于如何将两个有一定高差的地块在物质上和视觉上完美地整合。

设计师首先需要解决的一大技术难题就是如何在视觉上保持住宅与新地块之间的开放性（新地块将是私家花园的核心区域），而且使新地块与原有住宅之间具有明显的关联。设计师面临的另一大难题则是运用何种建筑元素连接起住宅与新地块，这也是原有设计方案的一部分。为了达到预期的设计效果，设计师在住宅后方设计了一道直线形绿廊与后花园相通。支撑绿廊的白色圆柱再现了原设计方案中摆放在正门的圆柱，绿廊的屋顶由木梁制成，可以为藤本植物和其他植物提供生长空间。此处空间可作为高架的直线形花房，其屋顶覆盖的一层绿色隔板可以保持热量，帮助植物快速生长，从而在绿廊的顶部形成一个枝叶葳蕤的绿色顶棚。

绿廊的一侧紧紧围绕着客房及为主宅提供服务的新设施，绿廊的另一侧朝向中央人造草坪；通道末端连接着错落有致的花园，这座花园占据了第二处场地的大部分区域，一道石墙将其与中央花园分隔开。石墙、泳池的边缘以及所有新的铺装表面都采用红砂岩构筑而成，以符合住宅入口处以及城市的传统色调。圣保罗乡间随处可见的巴西石在 20 世纪 30、40 年代被广泛应用于街道和户外路面的铺装。

紧邻石墙的游泳池是这座错落有致的花园与中央草坪的另一个连接点。泳池的一侧呈曲线状，与石墙、热带风情的花卉及灌木走向一致，另一侧则取直，池壁贴以绿色瓷砖，看上去就像是绿色草坪的自然延伸。游泳池的后面，翻过几级比较陡峭的台阶后，便能看

到错落有致的花园——其密集的植被设计包含了亚热带树种、棕榈、灌木、地被植物以及草本植物，所有的这一切搭建起一座天然屏障，保证了住宅的私密性。

玻璃住宅

该项目是对住宅各个部分边角"剩余"空间的利用和设计。设计对房屋的一部分进行了翻新，住宅南面有一处狭长的空地被两家共用的一堵墙围绕；与住宅翻新的侧厅垂直正交的是一个用玻璃构筑的新建筑物，与花园俨然一体。以建筑物为分界线，将花园分成两部分——东侧冷清毫无生气，西侧则是游泳池和平台。

一条蜿蜒的小溪将南面的后花园与东面的花园之间的视线连接起来，看上去就像是流经花园和住宅的天然水道。小溪沿岸是密集的植被，两侧的花园布局截然不同。后花园的水道一侧种植着喜阴植物和地被植物，另一侧则是铺筑好的步行区；穿越翻新的建筑物便可到达东面的花园，这里的小溪两侧种植着地被

植物、草本植物和棕榈树，形成一片郁郁葱葱的景象。

第三处空间是西面花园，也是整个项目中面积最大、最开放的空间。由于业主要求在这里设计泳池区域，所以设计师将花园一分为二，将较大的一块区域用来做游泳池。游泳池惟一的直线形侧边长 20m，其他三条边由三段连续的曲线构成，进入泳池中就犹如置身各种小花园的世界。木质人行桥是水面上惟一的景物，连接着种植在泳池边缘的草坪区域。草坪中央以花岗岩点缀并指向住宅公共区域的入口。泳池的一部分设计成室内游泳池，这样从室内也可清楚地看到花园和泳池水面。

植被设计保留了现存的古老树种和棕榈树，并添加了部分树种，如散尾葵、大鹤望兰、白鹤芋、阔叶山麦冬、宽叶山槟榔和银河粗肋草，以其作为花园的边界。该项目的结构、形状和色彩搭配美观大方，从而实现了和谐的景观效果。时隐时现的水景观将有助于营造一个休闲放松的周边环境。

1 玻璃住宅游泳池2

São Paulo, Brazil, is the largest city in South America and the fourth largest in the world, with more than 18 million people living in the metropolitan area and 11 million in the city proper. Although São Paulo, Portuguese for Saint Paul, was named by the Jesuits who founded the city in 1554, today's residents are descendants of immigrants from many other countries as well.

Since the mid-19th century, when the city began expanding rapidly as a result of coffee-growing on surrounding plantations, the capital of the state of São Paulo has grown continually and blossomed into Brazil's industrial and financial center. Traditionally, most of those successful industrialists settled in a neighborhood called Jardim Europa (Europe's Garden), whose streets are laid out in a very organic manner and whose large houses and gardens were designed in a very classic European style. The contrast between these classy and wealthy neighborhoods and the poverty-stricken image typical of the most central parts of the city is one of its most unfortunate characteristics.

However, as in any other of the world, these kinds of manicured neighborhoods add beauty to the city and the landscape design for every one of the private residences that constitute them become quite significant. In São Paulo, landscape designer Isabel Duprat is a well-known professional whose work focuses on these kinds of residential projects. Two of them, located in Jardim Europa, show creative designs, sometimes intimate and sometimes more open, where water is an important presence.

Yellow House

At the time Isabel Duprat was called to make a proposal, the owner of the house had purchased a second lot, adjacent to the one of the house, adding some complexity to the existing conditions of the site. The level of the new addition was more than one meter higher than that of the original lot, for this reason the biggest challenge of this project turned out to be the integration, both physical and visual, between the two.

One of the most important goals of the designer was to preserve open vistas from the house to the new lot, which would become a focal area of the garden, and an important visual connection from this new space toward the building. On the other hand, another main issue was the creation of an architectural element that would act as the connecting component between the two areas and which would appear as part of the original design.

In order to respond to this necessity, Duprat designed a linear pergola which runs along the back facade of the house, reaching the back garden; the white columns that make this pergola repeat the original design of the ones that belong to the original facades, and the roof is made of wooden beams on top of which vines and other plants grow profusely. This space acts as an elevated linear greenhouse because it is covered with a glass panel that retain the heat and helps these plants to grow faster, thus creating a dense green canopy on top of the pathway.

On one side, this pathway is enclosed by the guest rooms and some other new spaces that serve the main house and, on the opposite side, it opens up toward a central meadow-like area; the end of the walkway coincides with the elevated garden, which was created by taking over most of the second lot and which is differentiated from the central garden by a stone retaining wall. This wall, as well as the pool edges and all new floor surfaces, were built with red sandstone to match that historically used for this area of the city and parts of the entrance of this house. This Brazilian stone, natural to São Paulo's countryside, was frequently used for street and outdoor paving in the 1930s and 1940s.

Right next to this wall, a swimming pool establishes another point of contact between the elevated garden and the central widespread plane of lawn. Outlined by one curvilinear side, coinciding with the wall and a tropical-looking border of flowers and shrubs, and an opposite straight side, this pool was finished with green tiling

which makes it look almost as a natural extension of the lawn. Behind the pool, and some steps higher, the elevated garden presents a very dense planting plan based on mostly subtropical trees, palms, shrubs, groundcovers and herbaceous species, all of which generate a natural screen that provides privacy from the adjacent street and neighbors.

Glass House

This project is based on the use and design of the "leftover" spaces that remain vacant on the sides of the different components of the house. Part of the house was renovated and extends along the south linear side of the lot, leaving an empty narrow strip which is enclosed by one of the party walls; added perpendicular to the renovated wing is a new construction made of glass, which visually integrates with the garden. On both sides of this new addition, the garden is laid out as both a passive and an active area, respectively; the east side becomes the passive one and the west side incorporates the swimming pool and its decks.

The visual connection between the south back garden and the east garden is achieved by means of an artificial meandering stream that appears in both spaces, flowing at ground level, and seems to naturally go across the lot and the house. Alongside this stream spreads a dense planting

plan whose lay out is different in both gardens. In the back garden, the water line is framed only on one side, with shade species and groundcovers, and on the opposite side offers a paved walking area; in the east garden, which can be reached by going through the renovated building, the stream is framed at both sides by groundcovers, herbaceous and palms, thus generating a very luxuriant composition.

The third space or west garden is the largest and more open. Because one of the requests which the owner had was a 20 meter-long swimming area, the designer divided the space of the garden and took the largest side to outline a swimming pool. This 20 meter side is the only straight edge of the pool; the opposite edge is developed as a series of three continuous curves which seem to enter the water as different small gardens. A wooden footbridge is the only

piece that interrupts the aquatic surface and connects with the rest of the garden, which is a lawn area only planted on its edges and dotted with white granite pieces that mark the entrance to the house's public areas. Part of the swimming pool becomes an indoor pool, where the view of the garden and water is especially significant from the house.

The planting plan incorporated some of the existing old trees and palms and added some others to outline the limits of the garden. Textures, shapes and some color combine very soberly in this project and offer a warm and consistent image. Some of the species are Chrysalidocarpus lutescens, Strelitzia augusta, Spathyfilum canaefolium, Liriope muscar, Pinanga kuhlii and Aglaonema crispum. The presence of water, sometimes subtle and sometimes obvious, helps to create a very refreshing and relaxing ambiance.

1~4 玻璃住宅入口

海特经典住宅

Classic Residence by Hyatt

撰文 / 图片提供：SWA 集团　　翻译：张晶

该项目是一处服务型社区，毗邻著名的斯坦福大学。这里曾经是一家儿童医院，于1989年洛克庞马大地震后关闭。如今，这里已经成为了退休人员提供优质服务的景观化住宅区，原有的一些珍贵的自然资源、历史资源及配套设施也都得到了相应的保护和修缮。

该项目为住户提供了几种可供选择的健康生活方式——独立生活、协助生活和专业护理。全社区占地89 031m²，总体建筑面积65 032m²。该项目的设计规划主要从以下两点出发：解决不同年龄段老年住户的各种保健和活动需求；保护和科学管理现有的珍稀树种、历史古迹、考古资源和河岸走廊。

在将这块"棕地"重建为设施完备的服务型社区的过程中，设计师对环境问题给予了充分考虑。旧的建筑物被小心拆除，停车场修在了地下，原有的数百棵树木、稀有的当地植被以及一些濒临灭绝的野生动物的栖息地全部予以保留，并在其周围建起了一些新楼。为了保护当地植被，设计师从河岸边移栽了一些植物幼苗，这些耐旱的植物成活后不需要太多的养护。

社区还为老年人提供了丰富多彩的休闲娱乐活动——可以穿过马路到对面的斯坦福购物中心购物，也可以到隔壁的"麦当劳大叔""儿童健康理事会"等慈善机构做志愿者。与社区仅隔一块公共绿地的是斯坦福西部公寓和一家幼儿园，这里一些老年住户的家人就住在那里。

为了保护文物古迹、珍稀树种和其他当地植被，设计师在新的建筑物与沙山路和圣弗朗西斯基托河岸走廊之间留出了充足的空间，所有建筑物和道路的设计都尽量避免对其他树木造成伤害。最终的建筑结构布局既保留了原有树木，又造就了一系列各具特色却又彼此联通的景观，住户可凭借这些景观来区分各自

的单元楼，人们在此散步、休闲和社交的同时可以得到极大的视觉享受。

该项目的特别之处是保留了原址的历史古迹，如坐落于河岸走廊边的利兰·斯坦福驿车库（Carriage House）就是其中的一处，设计师对这里没有做任何开发，在原来的车库门前还专门设立了说明其本来用途的标志牌。如今，这所建筑和圣弗朗西斯基托河岸边的其他古迹都已经成为了历史的最佳见证，其他地方纵横交错的低矮石墙亦能唤起人们对农场旧貌的美好回忆。

户外空间设计包括休闲小径、花园、庭院和坐位区，住户们可以有更多的机会互动、交流。依据邻近建筑物的不同功能，每个花园还被赋予了不同的主题。在蔬菜园里，住户可以培育自己喜欢的蔬菜和鲜花，同时，为方便坐在轮椅上的人及其他行动不便者都能够参与，设计师还特意设计了加高的种植床。总之，这里为住户提供了高品质的花园式生活，住户们在此获得了视觉上的享受和情感上的极大满足。

Classic Residence by Hyatt is a continuing-care community located near Stanford University. Not only does the project provide a quality, landscape-focused facility for its senior residents, but it also preserves and enhances valuable natural and historical resources and amenities, replacing an old children's hospital that has been closed due to the Loma Prieta earthquake in 1989.

Classic Residence by Hyatt offers several health care options for residents, including independent living, assisted living, and skilled nursing. Facilities total 700,000 square feet within the 22-acre parcel. The site plan evolved from two core objectives: address the needs of senior residents at various stages of health and activity levels, and protect and maintain existing specimen trees, historic structures,

archeological resources, and the riparian corridor.

Careful environmental considerations were taken in transforming the existing brownfield site into this well-programmed continuing-care community. After the old building was carefully removed, a parking structure was built underground and the new building was designed around the hundreds of existing trees and the environmentally sensitive rare native plants and wildlife habitat, which included some endangered species. Drought-tolerant native plants were propagated from original seedlings found along the creek and then planted into the site to preserve native plant life, which requires minimal maintenance once established.

The community context offers a variety of recreational

activities for seniors that are within walking distance, including shopping at the Stanford Shopping Center across the street and volunteering at the Ronald McDonald House and Children's Health Council next door. Across a shared "Village Green" are the Stanford West Apartments and preschool, where families of some of the residents reside.

The SWA Group worked closely with the project architect in developing the initial site plan. Buildings are located with generous setbacks along Sand Hill Road and the creek corridor for the preservation of artifacts, heritage trees, and other native plantings. Throughout the proposed development, structures and paved access routes were designed to avoid all other existing trees. The resulting

building configuration not only saved trees, but created a series of unique, interconnected garden rooms. The gardens are now visual clues for residents to identify their homes, while also providing visually enjoyable places for walking, resting, reflecting, and socializing,

A unique opportunity of this project included the site's historic and archeological resources. One of the historic structures is the Leland Stanford's Carriage House located along the riparian corridor. All development was avoided in this area and a marker was erected to dedicate the structure and describe its original use. This historic building, along with other archeological sites, provides an interpretive trail that continues along San Francisquito Creek. Elsewhere

on the property, low stone walls arranged in rectilinear patterns recall the foundations of former farm buildings.

Outdoor spaces were designed to create a strong social community that encourages residents to take advantage of outdoor trails, gardens, courts, and seating areas. The design of each garden room has a distinct theme and use, according to the adjacent building function and desired activity. Residents can also plant their own food and other flowers in the edible garden, which was designed with raised planter beds so people in wheelchairs and with limited mobility can participate. Together, these elements provide a high-quality garden setting to be enjoyed by the residents visually, functionally, and emotionally.

阳光海岸

Sunny Bay

撰文 / 图片提供：ECOLAND 易兰

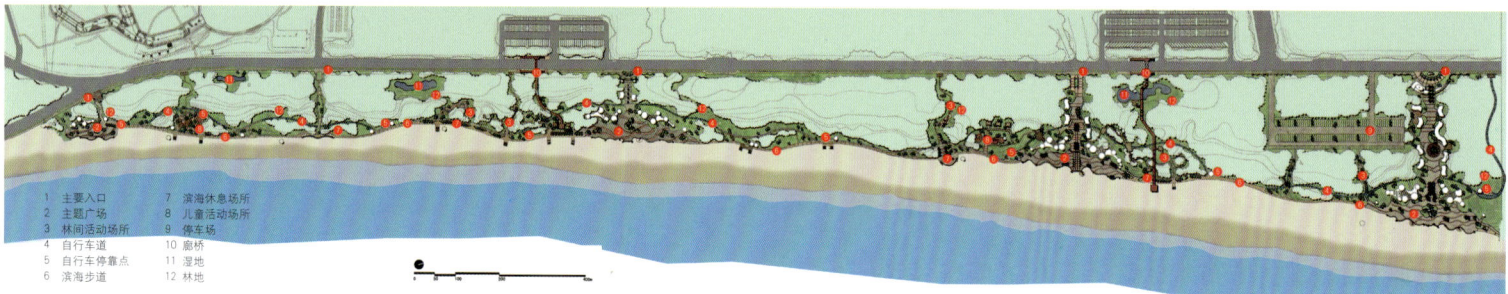

1 主要入口
2 主题广场
3 林间活动场所
4 自行车道
5 自行车停靠点
6 滨海步道
7 滨海休息场所
8 儿童活动场所
9 停车场
10 廊桥
11 湿地
12 林地

该项目位于山东省日照市碧海路以东，紧靠黄海西岸，海岸线长约 2700 米。场地南侧为帆船比赛基地和万平口水上运动公园，西侧为奥林匹克水上公园，北侧为滩涂湿地。场地东西向依次为城市界面—碧海路—较低的海滩绿化带—较高的沙坡高地—较低的沙滩—海洋界面，是典型的起伏的海滨沙丘地形地貌，现有植被以黑松林为主，植物种类不丰富且服务设施不完善。

该项目的规划遵循了生态优先原则，在保育的基础上对海洋、海岸、植被等自然资源进行再利用，充分考虑对生态资源如光能、风能、海洋潮汐等自然优势的利用，突出体现自然性、地域性、人文性、休闲性以及时代感，着重处理海与海滩、滩与场地、场地与设施、休闲观景与市政交通等关系。

整体场地包括几个主要区域：四个沿海岸分布的主题广场（领航广场、松涛广场、星海广场和乘风广场）；餐饮服务区；提供休闲步道和休憩场所等静态林下空间；提供迷宫、儿童游乐设施等动态林下空间；提供更衣、餐饮、水上中心和相关服务设施的中心服务区。木栈道和自行车道衔接了整体交通流线，沿途分布了名为"会海""望海""观海""游海""泛海""悦海""怀海"的多个休息点。

由于常年受海风影响，树木的生长速度极为缓慢，因此也尤为珍贵。由此，除原有的黑松林外，对重要地段受到破坏的植被，以种植恢复为主，同时选择耐盐碱、抗海潮风的乡土植物，合理配置进一步补充；采用三层复合式绿化结构，结合林带现状，形成近、中、远的植物景观层次，使场地植被界面更为丰富。同时，为了更好地保护植被，设计师在规划时将建筑散落在原有建筑基地和广场上，部分建筑与植被相交时建筑要进行避让，景观木道等也采用架空抬高的方式，使植被能保持自然生长。此外，波浪形挡土墙及其他景观设施不仅提供了休憩功能，也起到第二层沙坝的作用，可防止风沙对建筑的侵蚀，阻隔沙子向内陆的蔓延。

On the west coast of Huang Sea, east of Blue Sea Road in Rizhao City Shandong Province, is Sunny Bay, which has a coastline of about 2,700 meters. On the south side of the site are sail boat game base and Wankou Aquatic Sporting Park, to the west is the Olympic Water Park, and to the north are wetlands. The site is a typical undulating coastal dunes topography. Running east to west are an urban area, Blue Sea Road, the lower beach green belt, the higher sand hill and highland, and the lower beach, and finally the ocean. The existing vegetation consists mostly of black pine. The plants are unevenly distributed and sparse. The facilities there are also inadequate.

In keeping with the principles of ecological preservation, the planning incorporated conserved marine, coastal, vegetation and other natural resources, also fully considering making use of the ecological resources such as solar energy, wind energy, ocean tides and other natural advantages. There was a stress on reflecting natural, regional, cultural, and leisure with a modern aesthetic, as well as a focus on solving the relationship of sea and the beach, beach and site, field and facilities, recreational viewing and municipal traffic.

The whole site has several key areas: the four theme squares of Lianghang Square, Songtao Square, Xinghai Square and Chengfeng Square along the coast, and food service area, the glade providing recreation and open spaces for other functions, such as maze and children's play facilities, the central service area for bathhouse, restaurants, aquatic center and related service facilities. Boardwalk and bike paths linked the overall traffic roads, along which distributed number of break stations named Huihai, Wanghai, Guanhai and Youhai, Fanhai, Yuehai, Huaihai.

Due to the effects of perennial sea breezes, the growth rates of trees there is very slow, making the preservation and care of all plant material a top priority. For the destructed vegetation, the first step is restoration, meanwhile rationally plants that can withstand the salinity, wind and tide, and complement the native vegetation. These plants create a three-layer composite green structure combined with the existing tree belt to build different levels of landscape plants, making the site vegetation more abundant. To protect the original vegetation, the designer placed the buildings within the footprints of the previous buildings and the existing squares, so that no new buildings impacted the existing vegetation. In addition, the landscape boardwalk has been raised to maintain the natural growth of vegetation there. And the wave-type retaining wall and other facilities not only have a recreational function, but also played the role of the second layer of sand bar to prevent wind erosion of the building and block the spread of sand.

池畔新居 —— 翻修后的汉普顿庄园

Renewed Residence by Pond — Reconfigured Hampton Residence

撰文 / 图片提供：Sawyer/Berson Architecture & Landscape Architecture, LLP　　翻译：孙静静

随着汉普顿现代景观的发展，景观设计师对这座具有殖民地建筑风格的住宅进行了翻修。建筑与景观简洁的线条、协调的结构，共同铸就了这个和谐、典雅的佳作，使不同的景观元素既能相互顺承，又能保持各自独特的风韵。加强室内外连接与过渡是该设计中不可或缺的组成部分，使客户可以更舒适地享受这个避暑胜地的美景。

尽管该住宅所处的位置没有变化，但设计师对景观进行了全方位的改造。前方车道被迁移后，一片优美的草坪柔和地铺展在具有殖民地建筑风格的崭新门廊与前门之间；客厅和餐厅重新安装了凸窗，增大了廊柱空间，从里面向外望，人们的视线可以直接穿过草坪看到前方的垂钓池塘；住宅周围栽种着黄杨、吊钟花及玉簪，宛如起伏的波浪，巧妙地形成了可以遮蔽整个住宅的高大茂盛的榆树的装饰底围。草坪的一边紧临由木绣球组成的灌木花坛；另一边是与道路呈一条直线的，由玫瑰、杨梅、雏菊等组成的混合花坛，花坛四周都围有铁栅栏。

住宅的后方全部被改建成一系列相互交错、呈直线排列的房间，并且在外形和功能上相互衔接，形成统一的整体。不同的花园之间用女贞和侧柏组成的树篱、干燥的挡土墙以及草坪等隔开，四季常青的灌木花坛为不同的景观空间增添了协调性。灰色的砾石车道径直通往一个方形的机动车停车场，车道的纵向是由女贞树桩铺成，横向则是由灰色的混凝土路面砖铺成。这种路面砖还广泛使用在水池和阳台等的修砌中，使得项目整体的色彩格局一致、简洁。新整修的车库坐落于停车场的一端，它们构成了厨房后花园与农场的主要布局。

该项目整体布局呈一条规整的直线，当人们经过停车场，穿过白色的板条大门，来到泳池附近，进入凉亭时，这一直线形特征便映入眼帘。地面由灰色混凝土路面砖铺就而成，缝隙中生长着的许多青草巧妙地将硬景观与草坪衔接起来。坐落在远处的矩形水池映照出房子的倒影，也突显出后院的长度。水池的中央是一座简洁、典雅、垂着紫藤的凉亭，为泳池、温

总平面图

189

泉和住宅之间的过渡提供了阴凉的通道，装有灰色顶棚的方形温泉疗养区也是水池的重要组成部分。绿色的草坪一直延伸到规整的花坛边界，干燥的挡土墙上是侧柏树篱，支撑着由山茱萸、吊钟花、木绣球和冬青等组成的花坛，下面则栽有玉簪和常春藤等。

新建的阳台上设有户外起居室与餐厅，沿草坪向上延伸，地面沿用了灰色的、缝隙中长有青草的路面砖。阳台被牢牢地安置在呈"L"形的住宅中，另一端则固定在新建的灰色壁炉上，这个壁炉也是户外起居室的焦点，枝条蔓延的榆树恰好为起居室和餐厅提供了一个"天棚"。从厨房、门廊和客厅都能通往阳台，并且可将其顶棚延长，完全遮蔽这三个场所。到了夏季，阳台真正成为户外活动的好去处。冬青和黄杨犹如一条丝带，围绕在阳台的周围，几盆淡雅的雏菊和白色的新几内亚凤仙盆景点缀在休闲设施旁边。

月亮花园位于阳台后面，可以直接从主卧室走过去，里面的墙壁也是用灰色水泥砌成的。随意铺设的粗石路面上，补充了一些常绿植物，如黄杨、冬青、侧柏、卫矛等，粗石路面直接通向户外的喷泉和水池，而所有这些景观元素人们都可以从卧室直接看到。

前后庭院的过渡区是由一个蓝白交织的多年生花卉群形成的，四周环绕着女贞树篱，一堵白色的双层板条栅栏将花园与前庭院的绿色植栽隔开。

1　躺椅供主人在泡完温泉之后享受日光浴
2　被顶棚般的榆树遮蔽的主阳台和户外起居室
3　位于主阳台上的户外起居室
4　从水池望向主阳台的风景

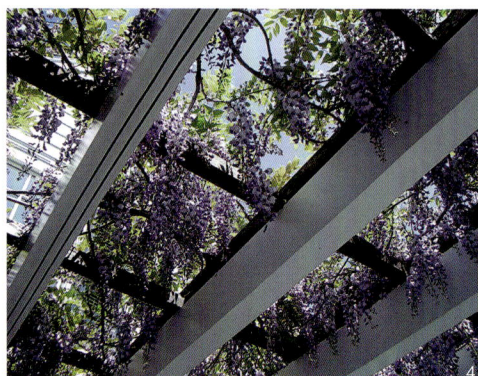

1 门廊休息区

2 从路面砖的缝隙长出的青草犹如地毯图案，
将户外变成名副其实的"起居室"

3、4 简洁典雅、垂着紫藤的凉亭

5、6 车道周围的植栽 1

This Hamptons contemporary landscape accompanies a Colonial Revival house whose substantial renovation was also done by the landscape architect. The lines of the house and landscape were simplified and organized to work in concert, yielding a balanced and gracious composition that allows the different spaces to flow one to the next but retain their own character. Enhancing the interior to exterior relationships and transitions was an integral design component, increasing the client's enjoyment of this beautiful summer retreat.

Although the footprint of the house remained almost the same, the rest of the property was entirely reconfigured. At the front, the driveway was moved to allow for a spreading, gracious lawn that gently swells to the new Colonial Revival porch and front door. The living and dining rooms with new bay windows and larger porch column openings look out across the green to Hook Pond below. The house is edged with a simplified palate of boxwood, enkianthus, and hosta in undulating waves that provide a subtle base for the magnificent Siberian elms that frame the house. The lawn is bounded by shrub borders of hydrangea to one side, and a mixed border of shrub roses, bayberry, and Montauk daisies that aligns with the road, all wrapped by a split rail fence.

The rear of the property was completely reorganized as a series of interlocking rectilinear rooms that relate to one another in form and function and present a unified whole. Privet and arborvitae hedges dry stacked and stucco retaining walls, and panels of lawn define the different garden spaces. The primarily evergreen shrub borders also lend unity to the different spaces. The gray gravel driveway gives on to a square motor court that is framed vertically by privet hedges and horizontally in gray colored concrete pavers. These pavers are used throughout the project for the pool coping and terrace paving, unifying and simplifying the overall color scheme. The renovated garage/pool house anchors one end of the parking court, and provides structure to the kitchen garden and utility farm at the rear of the property.

Careful alignments of the parts are apparent as one passes from the parking court through the white picket gate to the pool area and pool house pergola. The field of gray concrete pavers with grass joints provides a transition from the hardscape to lawn. The length of the rectangular pool

mirrors the long line of the house and accentuates the length of the backyard, anchored at the far end by a massive elm. Centered on the pool, the simple classical pergola adorned with wisteria takes advantage of the pool house/garage's exterior wall, and creates a shaded transition between the pool and spa and the building. A square spa banded with the gray coping is integral to the pool. The green expanse of lawn is bounded by the restrained border planting. A hedge of arborvitae set above a dry stacked retaining wall backs up the white flowering border of kousa dogwood, enkianthus, tardiva hydrangea, and inkberry. Hostas, vinca, and heuchera spread below.

The new terrace with outdoor living and dining rooms steps up from the lawn and retains a carpet of gray pavers with grass joints. The terrace is firmly nestled into the "L" of the house and is anchored on the other end by the new stucco fireplace, the focal point of the outdoor living room. The spreading limbs of the Siberian elm provide a suitable

1   车道周围的植栽 2
2   雨后的草坪
3   就餐区

canopy for the dining and living "rooms". The kitchen, new gallery, and living room all give access to the terrace, and the shed roofline was extended across all three rooms to unify the elevation. In the summer, the terrace truly becomes a destination for outside living. A necklace of inkberry and boxwood frames the edge of the terrace, and simple pots with daisies and white New Guinea impatiens punctuate the furniture groupings.

The private moon garden with its gray stucco walls sits behind the terrace and is stepped down to the level of the master suite. A complement of evergreen plants of boxwood, Japanese holly and arborvitae, and tiny Kewensis euonymus that ease out into the random fieldstone paving lead to the fountain and pool set into the structure of the outdoor fireplace, all seen from the bedroom.

At the far side of the house, the transition between the back and front yards is mediated by a blue and white toned perennial garden framed in a privet hedge. A white double picket fence separates the garden from the green expanse of the front yard.

东汉普顿住宅区

Residence in East Hampton

撰文 / 图片提供：Sawyer/Berson Architecture & Landscape Architecture, LLP　　翻译：傅冬

要为这栋位于东汉普顿的 20 世纪 90 年代的海滨住宅设计一处具有典型的巴厘岛风情的泳池和凉亭，需要设计师在这两处设施之间创造一种和谐的比照。位于泳池一侧的凉亭营造出一个私密的空间，而泳池的边缘逐渐消失，突出了海平面，令人浮想联翩，不禁想沿着蜿蜒的木板路漫步其中。

这个令人神往的异域天堂由一处泳池环绕，人们可以在此晒日光浴、用餐和散步，既方便休憩又可与朋友小聚。它将一个处于秀美风景中的海滨住宅打造成一个充满惬意的空间。新的暖色调的重蚁木露台和红木凉亭模糊了室内外的界线，延伸出去的与海湾相连的厨房、起居室和主卧室完全向露台敞开，使空间具有连续性，营造出空间的流动感和循环感。凉亭被镂空的屏风所环绕，将人们的视线引向前方的泳池，令人陶醉于一片旖旎的海洋风光之中。围绕壁炉放置的具有传统设计风格的沙发和垫脚凳更提升了房间内部的品质。凉亭的总体设计风格源于亚洲风格，一连串独立结构的凉亭位于露台的角落，营造出一种庭院效果。红色香柏屏风将凉亭和房屋环绕其中，营造了一处私密空间，将其与周围的住宅和停车场分开，使停车场和房屋前被一片以黑色和樱桃红色为主色调的水生植物所包围。穿过高大的镂空大门来到甲板上，人们立刻会有一种进入了具有迷人海洋风情的院落的惊喜。泳池两侧茂盛的季节性草木以及具有现代感的锌质容器进一步渲染了热带风情，仿佛从大西洋岸边来到了温暖的加勒比海岸。

位于露台前方的植被地带是从充满异域风情的植物到敏感的沙丘谷地的植物的生态过渡，并再一次强调了居民对私密空间和广阔的沙丘风景的偏爱。德克萨斯石灰岩质地的台阶沿着梯状平台铺设，逐渐变成石质台阶，饱经自然侵蚀的碎石镶嵌在由苦艾和蓬乱的百里香组成的地被植物之间，点缀在下方的平台上，这又为泳池泄洪道所在的、布满月桂树果实、李子和棠棣的斜坡提供了另一处休闲之所。沿着陡峭的斜坡向下，穿过草坪，就来到建有一长串台阶的石质码头。穿过位于草丛前端的镂空大门，踏上通往甲板的小路，人们可以在甲板上观赏夜景，璀璨的灯光照亮了通往

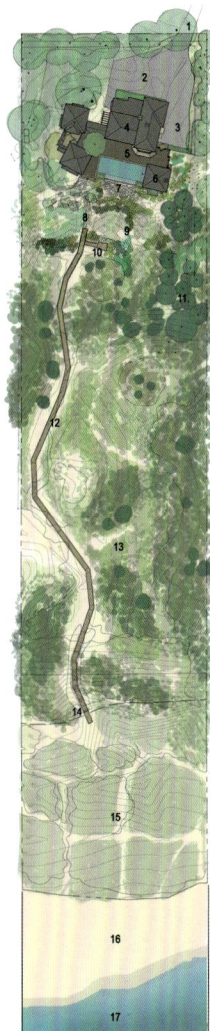
规划平面图

1. 车道
2. 汽车旅馆
3. 停车场
4. 现有房屋
5. 泳池与露台
6. 凉亭
7. 位于低处的露台
8. 楼梯与大门
9. 本地植被
10. 甲板
11. 常绿与落叶植物的过渡带
12. 木板路
13. 山丘居所
14. 位于步行道末端的大门
15. 海滩植被
16. 海滩
17. 大西洋

甲板的小路。

蜿蜒的木板小路穿过沙丘谷地,石楠、髯缀熊果和海滩草点缀着四周。走在木板路上的人们可以看到沙丘的顶部,而通向那里的道路则被隐藏在沙丘之间的低凹处。设计师还在裸露的沙地上种植了当地的草木以抵御木板路周围的风沙,台阶直至大门,大门与海滩的防雪栅栏相连,穿过沙丘地带,广阔的大西洋便跃然眼前。

1　柔和的灯光和点燃的炉火使人产生一种温暖的感觉

2　具有巴厘岛风情的凉亭

3　设置了舒适休闲躺椅的泳池

Designing a Balinese inspired pool terrace and pavilion as an extension of an existing 1990's beach house in East Hampton required creating a harmonious juxtaposition of two worlds. The pavilion flanks the pool terrace to create a private, intimate space, while the vanishing edge pool reiterates the plane of the sea and draws attention to the vista beyond, encouraging a walk along the meandering boardwalk through the delicate landscape of the Amagansett Double Dunes.

The exotic haven desired by the clients encompassed a pool, sunbathing, dining, and lounging, with protection from sun and cold, for both themselves and for entertaining. It transformed an existing beach house with free-form deck situated at a spectacular site into an extraordinary and unexpected delight. The new warm-toned ipe terrace and mahogany pavilion blur the distinction between indoor and outdoor as the kitchen with the renovated bay extension, the living room, and the master suite all open onto the new terrace providing a seamless flow of space and circulation. The pavilion, at once enclosed with roof and louvered screens behind, leads one to the pool steps and ocean vista beyond. Custom designed sofas and ottomans centered around the fireplace reinforce the interior quality of the space. The pavilion takes its massing from the Asian inspired house laid out as a series of separate structures and anchors the corner of the new terrace, providing a courtyard effect. The red cedar louvered screening wraps around the pavilion to the house, providing privacy from nearby neighbors and from the parking court. Leaving the parking court and front of house situated in a successional upland maritime forest of predominately black and pin cherry, one passes through the high louvered gate to the deck yielding the sudden surprise of entering a glamorous courtyard terrace dominated by the ocean vista. Lush seasonal plantings in planters flanking the pool and modern zinc pots reinforce the tropical feel, transforming the Atlantic setting to warmer Caribbean climes.

In the dune landscape below the terrace, the built and planted elements provide passage from the highly maintained exotic to the sensitive dune valley ecology, as well as reiterate the alternate desires for privacy and the experience of the sweeping breadth of the dune landscape. Texas limestone steps off the terrace lead to stone steps and a lower terrace of naturally weathered fieldstone set into ground cover of beach wormwood, bearberry, and woolly thyme. This provides a second entertaining area at the level of the pool spillway nestled into the restored slope of bayberry, beach plum, and shadblow. The steep slope down through the thicket cleared of invasive plant materials is managed by a string of ipe steps and stone landings. Passing through the louvered pool gate at the bottom of the thicket, one steps onto the boardwalk as the landscape opens up to the ipe clad moon deck for intimate nighttime stargazing. Flush mounted lights illuminate the boardwalk for moonlit strolls.

The silver ipe boardwalk winds through the dune valley landscape of shifting masses of beach heather, sandwort, bearberry, and beach grass. Placed along an existing sand path, it keeps the final dune crest always in sight, as the immediate way is hidden by the curves of the dune it hugs. Within months of construction, native plants filled in what had been bare sand, seeking the wind protection of the boardwalk structure. Finally arriving at the primary dune, the boardwalk climbs an existing pass with stairs ending at gate that physically and stylistically connects to the existing beach snow fencing, as one passes over the dune revealing the expanse of the Atlantic Ocean.

1　铺着传统绒布的躺椅沿着泳池摆放，逐渐消失的泳池边缘突出了前方大西洋平静的海面

2　凉亭中的沙发、垫脚凳和石桌为观赏独特的沙丘风景提供了一个绝妙的去处

3　内部设计成一个户外起居室的凉亭

洛杉矶新艺术中心区的"小村庄"—— 豪华Mura公寓

Little Village in the New Arts District of Downtown L.A — Mura, Luxury Condominium Complex

撰文：Barbara Pressman　　图片提供：Greg Epstein Photography　　翻译：刘建明

　　位于洛杉矶市区工业区旧址附近的全新奢华公寓——"小村庄"项目，已于2008年春季竣工。因为临近日本街，所以这个由普尔特房产公司（Pulte Homes）加州欧文分公司开发、共有190个单元的豪华公寓项目被命名为"Mura"——在日语里就是"小村庄"的意思。

　　由于该项目临近日本街且位于工业区旧址上，在这样的背景下，景观设计方案不仅包含了钢桥、钢梁和钢轨等景观元素，同时还不乏诸如竹子、石头喷泉和幕墙等具有亚洲特色的景观元素。

　　据该项目的景观设计师、总部位于加利福尼亚州安西诺的TGP景观设计事务所的合伙人罗布·普雷斯曼（Rob Pressman）称，此设计方案的美学特点即现代城市主题，受到了该项目所处的位置——洛杉矶市区新的艺术中心区的启发，该艺术中心区以具有强烈对比效果的结构、图案、角度和钢材构成的古老的、仓库式建筑物为特点。

庭院

在高于街区水平面两层楼的车库顶部位置，设计师为该项目设计了4个庭院。庭院之间用五颜六色的步行道相连接，为周围单调沉闷的楼体增添了几分暖意与色彩。步行道采用四种颜色的混凝土材料以某种重复的图案铺筑而成，形成了颇具现代风格、编织精致的"地毯"，并且巧妙地贯穿整个项目。

其中一处庭院是居民进行社交活动的户外起居室和餐饮区；另一处庭院设计为一座水池和温泉区，从这里可以观赏到独特的城市风景；余下两处以植物配置为特色的庭院使公寓各个单元的人们都能领略到其中诱人的景致，同时也为低层单元增加了私密性。

设计师运用了墙体、高差和其他景观元素来打造该项目的核心——占地面积为465 ㎡的户外起居室，其中紧邻的几座人造喷泉将起居室分隔成几个独立的空间，分别供居民就餐、烹饪以及尽情享受炉火的温暖。起居室的周围栽满了交错排列的竹子和罗汉松，并且用0.9m×2.4m的镀锌钢条格栅围护起来。

连接步行道和水池的是一处非常具有表现力的现代化钢质高架结构，其延伸部分宛如悬臂，营造出一片阴凉。温泉区位于低矮挡墙的附近，通过挡墙的玻璃窗可以看到街面，欣赏如画的城市风景。

项目周边外部景观设计

项目周边外部景观是指公寓正门与步行道之间的狭长地带，设计师有意创造出两种不同风格的景观：一种景观风格与建筑实体的规模有关；而另一种景观风格则更人性化，围绕临街的第一层公寓的露台来做文章。

通过种植竖直生长的常青植物，例如成年后可以长至12m～18m高的斑竹、雪松、加那利群岛松和红胶木，现在看来还是一片空白的公寓正面不久之后将隐没在一片翠绿之中。对于带露台的第一层公寓单元来说，在围栏和门廊之间栽种一些形体较小的花卉植物比较合适，例如紫荆和紫薇。

该项目最初是针对年长的日本买家而设计的，据统计，这个群体的人口数量在邻近区域中所占的比重最大。然而，最终有大约一半的公寓单元被韩国人买去，他们大都是为在附近南加州大学上学的孩子们而购置的。今天，在这处温暖而迷人的公共开放空间里，无论是壁炉旁还是池水畔，散布着三三两两的年轻人，正符合"小村庄"祥和宁静的氛围。

1　在温泉区可以领略到独特的城市风景

2　最大的庭院——户外起居室和餐饮区

场地平面图

2 步行道

Mura, a new luxury condominium complex in a burgeoning neighborhood of a formerly industrial section of downtown Los Angeles, was completed in the spring of 2008. Developed by Pulte Homes of Irvine, California, the 190-unit luxury development received its name - which means "little village" in Japanese - because it was situated immediately adjacent to Japan town.

This proximity to Japan town and the overall industrial context of the project inspired a landscape design palette for the project that includes steel bridges, beams and railings, as well as Asian elements such as bamboo, stone

fountains and walls.

According to Rob Pressman, partner in the landscape architecture firm TGP, Inc. Landscape Architects, of Encino, California that designed the landscape for this project, the design aesthetic is a contemporary urban motif derived from the project's location within the new arts district of downtown L.A, which is characterized by a mix of old, warehouse-type buildings composed of contrasting textures, patterns, angles and steel. The firm was engaged by Pulte Homes to design the common open spaces for the project as well as the exterior landscape surrounding the project.

Courtyards

Two stories above street level, atop the parking garage, TGP, Inc. designed four courtyards for the project.

The courtyards are connected by a colorful promenade that adds warmth and color to the muted color palette of the surrounding buildings. Paved with four colors of interlocking concrete pavers in a repeating pattern, the promenade creates a richly woven contemporary carpet that unfolds throughout the project, connecting the courtyards.

One courtyard was designed as an outdoor living room and dining room area where residents could socialize; a second courtyard features a pool and spa with spectacular city views; and two planted courtyards provide attractive views for the surrounding units as well as privacy for the units on the lower floors.

To create the 5,000-s.f. outdoor living room, which is the heart of the project, TGP employed walls and varying elevations and other design elements, including back-to-back fountains that create separate rooms for dining, cooking and enjoying the fire. The space is accentuated with alternating rows of bamboo and fern pines, and framed with three-foot by eight-foot, galvanized steel-bar grates.

At the pool area, a dramatic, contemporary steel overhead structure provides a link between the pool area and the promenade, as it cantilevers out into the pool area, creating a shaded sitting area beneath it. The spa is located adjacent to the parapet wall overlooking the street, and enjoys spectacular city views from a glass window in the parapet.

The Exterior Landscape Surrounding the Project

While the exterior landscape of the project occurs within a narrow strip between the building façade and the sidewalk, the challenge for TGP was to create two landscapes: one that relates to the scale of the building as a large entity, and a second, more human-scale landscape that relates to the patio ground-floor units along the exterior of the building and engages the street.

By planting vertical-growing evergreen materials such as giant timber bamboo, cedar trees, canary island pines and brisbane box that will eventually reach 40-60 feet tall where the building meets the ground, the now-blank building facades will eventually be softened. For the patio units, smaller-scale, flowering materials like Hong Kong orchid and crepe myrtle trees were planted between the security fence and the stoops.

Mura, which means "little village" in Japanese, was originally aimed at elderly Japanese buyers, the largest demographic in the neighboring area. Ultimately, however, around 50 percent of the units were acquired by Korean buyers for their children, who were attending the University of Southern California. The warm and engaging public open spaces at Mura, filled now with young residents socializing around the fire or at the pool, truly make it the "little village" that its name connotes.

1 庭院间的步行道和墙体
2 现代化钢质高架结构
3 具有现代感的钢桥
4 喷泉和树木处的灯光给人温馨的感觉
5 餐饮区
6 户外起居室的壁炉

似曾相识的体验 —— 金阳新世界入口广场

Culture Transformation — Jinyang New World Entry Plaza

撰文：安庚心　　图片提供：安庚心　庄迪　AECOM 中国区规划 + 设计（广州办公室）

艺术是对生活体验的一种表达，借由画作、雕塑、景观设计作品以及不同的艺术处理手法，将个人抽象的情绪、想法、概念、生活体验等转化成可观、可触、可游的实体。任何艺术创作都需要灵感的启发，无论是不同的地域特点、总体规划上的考虑，还是氛围的塑造，都可以成为创作的出发点。在该项目的景观设计中，设计师致力于将当地的生活体验，用简单、直接、令人印象深刻的设计手法表达和传递出来，希望人们能通过该项目的景观设计为当地历史悠久的本土文化所触动。

该项目位于贵阳市地理中心新世界金阳新区的南北主轴线上，占地面积为 12.69 万平方米，是一个新型的现代居住开发区，也是一处带有古朴庄重的韵味、自然与人工相结合的现代景观。

在该项目中，设计师以当地文化为灵感，充分结合当地的自然生态环境、衣、食、住、行等民俗风情及当地材料与颜色细节，综合提炼出其中的神韵，从而形成灵感来源，借助现代设计手法对当地的特色重新进行演绎，使人融入到大自然中，感受其中的氛围，给人以似曾相识的亲切体验。

结合当地著名的景点黄果树瀑布作为一种景观文脉，设计师创造出具有地方特色的水景元素。入口的水景虽方整抽象，但与植栽结合成为抽象的自然瀑布，形成自然的一种再现。由此产生竖向层次的变化，并与整体的景观设计达成微妙而和谐的平衡。桌，在中国传统中有团圆的寓意；涌泉，可静——如镜产生变幻的倒影，可动——产生源源不断的表情，这便是水景的表现力。大地艺术在此也有所体现，受到当地特色农业景观——梯田的启发，设计师设计了梯田式的大型水景，以此作为视觉的焦点，形成与都市区间的有效过渡。同时在水景中还有抽象化的河流与岛屿，形成丰富的景观层次和变化。

设计师除了因地制宜地将当地文化风俗作为灵感的出发点，在设计中也非常注重在细节上的考量及优化，匠心独具的细节让整个项目更富于文化内涵，带给人们更多的愉悦感。

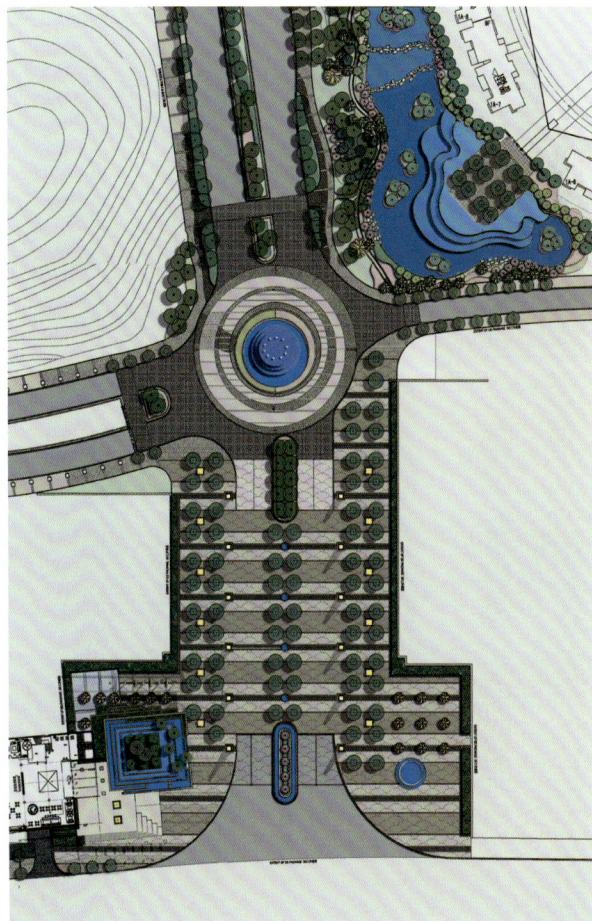

平面图

金阳新区的前广场与公园住宅区在设计中都运用了一些竖向元素，如玻璃体、灯柱和高大的乔木，以此营造出商业氛围。考虑到需要营造出日夜交替的景观变化以及当地阴霾的天气，设计中特意采用强烈的红色、橙色系列，以达到营造商业氛围的目的。玻璃雕塑体中红色与橙色的运用，反映了当地夜间集市的特色，以艺术的手法重新诠释景观，同时，也强化了商业街的印象。玻璃体四面的图案相似却不雷同，人们能从中感受到景观的细腻变化。边长为2m的玻璃立方体可以产生纯净的空间感受，同时也具有灯具的功能，人们虽然不能走进立方体，但其以阵列的方式排列，可以产生一种方向性的导引，营造出热闹的氛围。10m高的地标性引导灯柱，更是功能、艺术与景观的高度融合，其设计灵感取自于古塔，并以现代的手法进行再创作。

优秀的景观能给人以联想的空间。敏感的设计手法、当代精神理念恰到好处地与当地特色相融合，既能引起共鸣，又能从共鸣中找到新鲜感，并借这种新的感触激发出更多、更新、更好和更远的思考，这正是一个成功的景观设计的意义所在，也是该项目景观设计的成功所在。

1、2　玻璃体与红色长椅相得益彰
3　灯柱及其倒影
4　鲜明的颜色强化了景观氛围

Through different media such as painting, sculpture and landscape architecture, art conceptualized the intangibles and turned them into the tangible objects. The creative process of art making requires a strong concept and often draws its inspiration from nature.Landscape architecture in particular, draws its inspiration from the existing site conditions, surrounding urban or rural context as well as the native culture of the place.

Jinyang New World is located in the Jinyang district, a newly planned CBD of the Jinyang city. The scope of work includes the entry display area and a streetscape, totaling 12.69 hectares. The contemporary landscape design of Jinyang New World's entry plaza is strongly influenced by the native culture, as well as the existing urban context. The landscape designer is inspired by the native architecture, artifacts, agriculture landscape and urban life style. Designer transformed the native elements and both urban and natural sceneries to series of contemporary landscape features. The drum tower, rice terraces, stone wall from the ancient village, and the native jewelries design become the source of the

inspiration for the contemporary landscape design.

The drum tower is transformed to the urban totem and used as lighting feature along the entry plaza. The form and shape of the rice terraces can be found at the water cascade at junction point. Local stone are used for the planter and retaining walls while the jewelries pattern is applied as paving patterns for the future commercial plaza. These visual native elements are been reinvented in a modern way by means of abstracted forms, shape and materials. Since the entry plaza also serves as access point to the future commercial areas, rows of colorful light boxes are used to reinforce the commercial ambience which resembles the Guiyang night market.

The entry plaza creates memorable experiences for both the local residents and the visitors. For the local residents, theses culturally inspired inventions unconsciously reminds them of their own past. As for the visitor, the plaza offers a refreshing experience which is truly unique of its kind.

1　等距排列的图腾灯柱与矮柱灯
2、3　跌水瀑布

新古典主义住宅 —— Grand Bourg公寓

New Classic Residence — Grand Bourg Building

撰文 / 图片提供：Jimena Martigoni　　翻译：赵玮

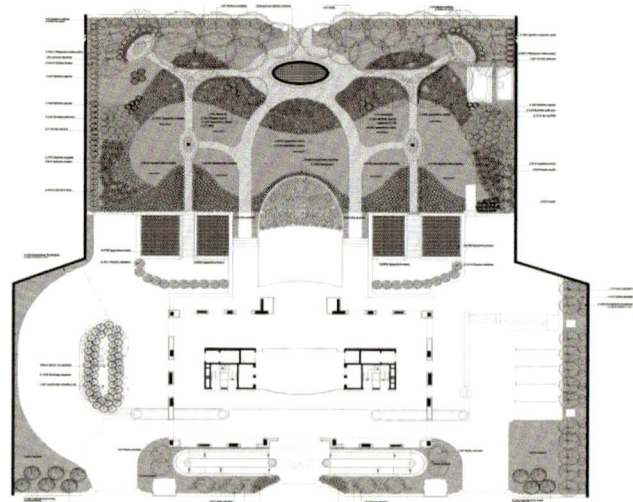

总平面图

　　该项目是在布宜诺斯艾利斯设计建成的第一批住宅公寓之一，几年前它的出现，还引发了一场新古典主义的风潮，其复古的法式建筑风格受到了全世界很多城市建筑的争相追随。

　　该项目远离市区，独享着周边优越的居住环境，在所有通往布宜诺斯艾利斯的路途中，这里的风景最为优美，视野开阔，四周密林环绕。该项目还紧临这座城市里最古老的一条街道，并且可以很便捷地到达新建成的拉美艺术博物馆 (Museo de Arte Latinoamericano de Buenos Aires or MALBA)。

　　博物馆的创建者不仅投资兴建了这座博物馆，还负责艺术品的收集，同时也是该项目的策划人。他邀请同一组建筑师来完成这两个项目，结果证明了这个决定是正确的——博物馆建筑的设计充满现代、前卫的时尚感；住宅项目的设计同样在很大程度上引领了目前房地产市场的流行趋势。

　　该项目的景观设计充分体现了设计的多元化，需要尽可能解决以下问题：要与建筑设计和总体规划相呼应；引入自然元素来软化僵硬的建筑轮廓；同时按比例缩小花园，营造出亲切宜人的舒适氛围。为了实现这些设计主旨，景观设计采用了规则的几何式布局，以呼应法式建筑的轮廓及风格，并吸收了一些古典别墅花园的设计元素，在植物配置上也尽量做到还原真实的自然环境效果。

　　由于场地条件的限制，花园的大部分只能建在地下停车场之上。因此，景观的规划理念随之调整为打造一个"绿色屋顶"，将高大的乔木配置在具有适宜种植的土壤的场地后侧。其余的场地，如楼梯、台阶、穿堂和它们之间的开放区域，则主要种植地被及灌木等低矮植物，来突出建筑的宏伟高大。

　　从屋子的窗户或者阳台向外眺望，那将是一种享受——五彩斑斓的花园，整齐规划的小路，一天当中随着时间变化而呈现不同的美景；中轴对称式的花园布局及合理的种植规划都使古典与现代元素巧妙地融合在一起，相映成趣，身在其中不禁使人乐不思蜀。

　　通往室内的台阶中间围合出一个半圆形的挡土墙，四周由矮灌木环绕。设计师在挡土墙的内侧设计了一个绿色缓坡，避免了视觉上的突兀；挡土墙的外侧种植了色彩鲜艳的小花，无论从哪个角度欣赏都像是经过精心修饰的花环。这片区域的下方是一个游泳池，旁边是通往下层空间的台阶。

挡土墙后面的花园以一片草地为中心，两侧各种植一小片花灌木丛，狭窄的小路以几何模纹的形式将绿化区域划分开来，在小路的末端有一个椭圆形小水池，这里便是花园的尽头。这样的围合与划分，使花园成为了视觉的焦点，两侧茂密的植被不仅提供了休闲纳凉的好去处，也围合出了更多宜人的私密小空间。蓝花楹、合欢树、刺桐、珊瑚树……在这些树下的长凳上即便是短暂的小坐，也能博取片刻的宁静与放松。一排梧桐树（法国梧桐，悬铃木属二球悬铃木种）像

围墙一样矗立在这片宁静空间的边界，密不透风的树冠在酷暑时节为居住区里的人们挡风遮阳的同时，也将闹市的喧嚣留给了外面的街道。

草坪四周环绕着萱草和百子莲，其绚烂的色彩夺人眼球。这些花的花期很长，可以从春天开始绽放，直至夏季结束。当鲜花盛开时，花园里的台阶就像镶嵌了花带，这些花带遍布于整个花园，引导着人们走向草坪。设计师塞西莉亚·默里说："我非常喜欢这些创意：用植物弱化阳台和台阶，让自然植被作为入口的延伸。"行走在无数的黄色萱草中，着实令人沉醉。

所有草坪均略高于周围的小路，原场地单一的地形经过处理变得高低错落，不仅形成了一处富有活力的空间，也增加了游园的趣味性。

院墙上种植了攀缘植物，如波士顿常青藤和玫瑰，另外，在紧挨着墙角的地方整齐地种植了一排冬青（冬青属）和地中海荚蒾等，同时还配植了一些蝴蝶灌木等。

建筑的正立面与一条街道相对，从街上望去，新古典主义的建筑轮廓显得更加突出。对称式的平面布局及植物配置与住宅建筑完美地结合。主入口两侧的柏树和大小不同的西谷椰子（苏铁）环抱于建筑两侧，使其具有了皇家园林的气派。

该项目的别墅样式经典，甚至连院落都有明显的复古特征，花园配合其特定的建筑特色呈现出强烈的新古典主义色彩，同时又以独特的现代方式与自然环境完美地结合，不失为一项成功的景观作品。

Grand Bourg is one of the first apartment's buildings in Buenos Aires which was designed, a few years ago, evoking a neo-classical style and following a French architecture particular retrospective that is nowadays affecting many cities in the world.

These kinds of buildings are built as exclusive residential towers located at exclusive areas of the city; in this case, the building is sited at one of the most picturesque avenues of Buenos Aires, particularly wide and framed by large trees, next to the most traditional neighborhoods of this city and, also, to one of the newest and now most important museums of Latin America, the Museum of Latin-American Art of Buenos Aires (Museo de Arte Latinoamericano de Buenos Aires or MALBA).

The entrepreneur and art collector who proposed and funded the construction of this museum is the same person behind the project for Grand Bourg and, for this last one, he called the same architects. However, as much as the museum is an example of contemporary and innovative architecture, the residential building is one that responds basically to current real-estate trends.

The landscape plan wants to adapt and respond to the general concepts and architectural design of this building but, at the same time, seeks to create a place of diversity where the offer of intimate spots scale down the size of the gardens and where the lines of nature help to "soften up" those of the architecture. In order to achieve this, the project is materialized as a formal and geometrical layout that accompanies the French lines of the architecture and incorporates certain components of the classical villa garden and, on the other hand, achieves a natural-looking space through the organic shapes of the planting plan.

One of the main constraints of the location was that most part of the garden area had to be built on top of the underground parking; for this reason, the landscape plan had to be thought out as a green roof and, consequently, all the trees had to be grouped at the rear of the site, where the ground floor is over natural soil. The rest of the area, which faces the stately stairs and terraces that connect the lobby of the building with the open spaces, is fully planted with herbaceous, shrubs and groundcovers that don't require excessive heights of soil to be planted.

Another relevant condition for the landscape layout was that it would be mostly appreciated from the windows and balconies of the apartments. This bird's-eye view had to be carefully taken into account to outline the system of paths and green areas of the garden which, altogether, would allow a consistent and attractive image for the residents. The symmetrical arrangement and axial structure of the garden and pathways respond to this situation and adapt a modern planting plan to a classical design where every one of the parts is related to the whole.

From the public terraces, the first element to be clearly made out is an inclined green plane that covers a semicircular retaining wall that otherwise would have looked as a large bare vertical one. In order to hold the earth that was incorporated here to model this plane, the designers had to use a system of geotextile cells onto which they planted a well number of groundcovers and a row of flowering small shrubs that crown this surface. Right at the bottom of this space lays a swimming pool, developed underneath the terraces.

Behind this plane, the garden is exhibited as a geometrically differentiated space with a central lawn area flanked by two adjacent smaller ones at both sides.

These green areas are limited by narrow stony paths that lead to the rear side of the plot and which converge into a small oval pool. This central focal point becomes an eye-catching element of the garden and announces the creation of a more intimate, shady space edged by large native trees. Underneath these species—namely Jacaranda mnimosifoia, Chorizia speciosa or floss-silk tree and Erythrina cristagallis or coral tree—resting spots only furnished with solitary benches invite to a moment of distraction and relax. Positioned as the last component of this sheltered area, a row of sycamores (Platanus acerifolia) create a dense green canopy which during the hottest days of summer provide a refreshing ambience and also create a natural screen that

conceal the gardens from the back street.

This composition establishes a deep contrast with the bright colors of the daylilies and agapanthus that massively cover the edges of the lawn areas, in the open. These clusters of plants that bloom as soon as the spring arrives and keep offering color all summer long are displayed as extensions of the terraces that get into the lawn areas and are perceived as stains that spread over the garden. "I very much like the idea of balconies and terraces overlooking another space and using plants as their natural extensions" says Cecilia Murray, one of the lead designers. In this manner, large masses of yellow daylilies define borders and seem to move back and forward between the terraces of the building and the garden below.

All these lawn areas are slightly elevated from the ground level, or that of the paths, eliminating the idea of a purely horizontal plane and offering a subtle elevation change that makes a more dynamic space.

For the party-walls, the landscape designers decided to use climbing species such as Boston ivy and roses, and for the foreground, right in front of them, they planted some formal parterres of Hollys (Ilex stevens) and Viburnum tinus and other more casual groups of Butterfly bush (Buddleias).

At the main facade of the building, which face a grand avenue and where the lines of the neo-classical architecture are stronger, the design team decided to use a symmetrical layout and plants that conceptually allude to villas and formal gardens. Cypresses that flank the main entrance and Sago palms (Cycas revoluta) of different sizes, which are aligned in front of the arches of the building, create a welcoming imperial plane especially thought to be appreciated from the street.

Altogether, the garden for the residents of this particular building has clear reminiscences of a classical composition, typical of a villa, and establishes strong references to those yards from the past. Yet, the natural approach with which it has been outlined and planted proposes an alternative vision and a modern answer to a neo-classical general configuration.

幸福之家

Baan Sansuk

撰文 / 图片提供：Pok Kobkongsanti (T.R.O.P.)　　翻译：董桂宏

泰国华欣的海水清澈湛蓝，海床上巨大的砾石清晰可见，也因此蜚声世界。该项目就坐落在风景如画的华欣海滨，是一个高档的住宅项目。

该项目的场地形状类似中国的面条，细细长长，一直延展着通向海滩。从住区门廊到海滩区的场地宽度由 10m ～ 12m 不等，这种狭长的建筑布局使得各个建筑面面相对，而住户也无法观赏到华欣海滨的美景。建筑师将两条建筑中间狭长的空地留给景观设计师进行填补，因此景观设计师的首要目标是为住户创造出独特的视觉享受。

景观设计师将华欣海滨的特色美景搬到了住户的前庭：该项目外围由长达 230m 的水池环绕，其中包括一系列小水池。第一座水池叫"迎宾池"，靠近大厅，约 60 厘米深，孩子们可以在这座巨大的浅水池中戏水；第二座水池叫"衔接池"，因为该项目的场地十分平坦，设计师试图通过在某些位置将池塘垫高以形成视觉高差，因此建造了逐步升高的台阶用以调整高差，"衔接池"也因此得名；第三座水池是SPA 池，设计师特意在池中设计了水力按摩浴缸和鼓泡床；最后一座水池面向海滩，并且没有设置边界，住户可以在此一边游泳，一边将华欣的海滨风光一览无余。

景观设计师基本以华欣当地的元素为主进行设计，如绿色的草坪、白色的砾石路和鸡蛋花树等，并以一种独特的方式对这些平常元素进行重组，营造出独一无二的住宅景观。

该项目的整体设计为一系列与水景交织排列的曲线道路，景观设计师特意选用曲线的形式以柔化建筑硬朗的线条。在某些特殊的地块，道路从场地的一侧转向另一侧，蜿蜒曲折，妙趣横生。虽然花园受到了场地地形的限制——又长又窄，但趣味性的设计使得住户获得了特殊的散步体验，步移景异，大大丰富了视觉体验。因此，该项目的设计获得了巨大的成功，并成为滨海住宅开发项目中的典范。

Baan Sansuk by Sansiri is a High End Exclusive Residential, located at the most popular Beach of Thailand, Hua Xin. Hua Xin is famous for its huge stone boulders in the clear sea water.

The Site is a very long strip of land, with a narrow Beach Access, similar to Chinese Flat Noodle. When we got a commission to design this project, the Architect already located 2 Long Rows of Buildings on both sides of the land plot. In the middle, they gave us a long narrow gap, with an average 10-12m width from the Lobby to the Beach area. Because of the lay out, now each building is facing another building. With no direct view of the Ocean, our first move is to create special view for our residents.

So we try to bring "Hua Xin" to their front yard, by placing 230m long pool throughout the site. The 230m long Pool is a series of smaller pools, connecting to one another. The first pool is a Welcome Pool, about 60cm deep. So kids can enjoy this huge shallow pool near the Lobby. The second pool is a Transition Pool. Because the site is very

flat, here we try to create level differences by raising some part of the pool up. The result is those raised water steps in the middle of this pool. The Third Pool is the Spa Pool. We strategically locate Jacuzzi and Bubbling Beds under the water here. Finally, we have located our Infinity Edge Pool at the Beach Front. Here our residents can enjoy a good lap swimming with the unobstructed view of Hua Xin Beach.

Basically, all design elements are the local elements in Hua Xin's existing elements. Green lawn, white gravels, plumeria trees are common in the area, however, we tried to arrange them in our own way.

Overall design is a series of Curvy pathway interlocking with the Aquascape. We intentionally use the curvy line to soften the hard edge of our buildings. The pathway will also switch from one side to another at some strategic points. This is to create more interesting experience walking through a long and narrow garden. As a result, the residents will change their views as they travel along the pathway. At the end, this project is a huge success and become the case-study for Beach Front Residential Developers.

NOVA住区庭院

NOVA Residential Courtyards

撰文 / 图片提供：mcgregor+partners　　翻译：武秀伟

1

该项目位于昔得兰的格林广场上，以维多利亚公园为背景，总建筑面积为 6265m²，分为两处庭院。在这两处庭院的四周建有 4 栋大厦，共 119 套公寓。其中，一处为中心庭院，其设计充分体现了维多利亚公园的开放式空间布局，为周围居民提供更多的活动空间，满足人们日益增长的娱乐需求；另一处为公共庭院，也是大厦的延续部分，它以线性的方式将中央空间分隔开。两处庭院均设计了鲜明、时尚的图案结构，人们无论站在地面还是阳台上都可以观赏到这些景观元素。

mcgregor+partners 事务所赢得了维多利亚公园的开发项目，并与 Turner and Associates、开发商 Waltcorp 合作。从整体规划到细部设计，这支久经考验的、具有创新精神的团队都发挥了重要的作用。

这两处庭院都设计成花园形式，在土丘状的草坪上种着纤细的桉树，周围是波浪状草地和沙砾铺设的小路，园内分布的草丛和石头暗示着此地曾经的生态地理环境。这些特征依据地区生物多样性理论，体现了现代设计风格的发展趋势。鹅卵形土丘增加了土壤的深度，有利于树木的生长和根茎的舒展。两处庭院与维多利亚公园及其他公共空间形成了一个内部循环系统，并与住宅区分隔开，构成一处自然隔离带。

庭院景观的设计遵循了环境设计原则，不仅关注视觉效果，还考虑到与环境的协调性，从而营造一个健康的生活环境。设计的首要目标就是在控制建筑预算的前提下，使建筑与景观的各个方面均达到生态的可持续性发展。植被也非常重要，这里的植被几乎囊括了澳洲本土的所有植物种类，而在建筑物的南边还种植了树蕨、棕榈和其他耐阴的本土植物品种。如今的 NOVA 已经十分完善，并成为了陆地动物群的栖息场所，一些鸟类也在此栖息，如雀类、蓝鹩鹩和澳洲喜鹊。低耗能的、可循环使用的材料也得到了充分利用，如悉尼沙岩和木材。

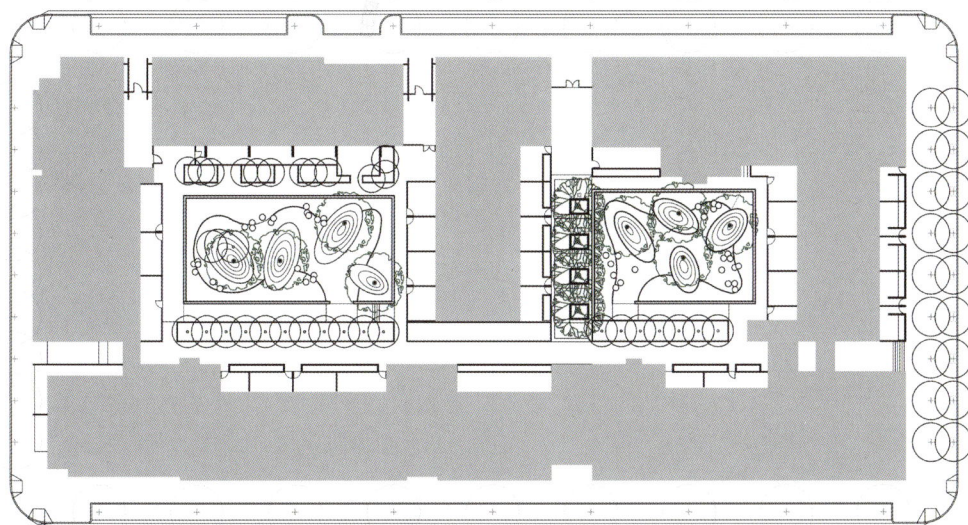

中央庭院的四周是一些住宅区域。考虑到每个公寓楼的结构，设计师分别设计了前庭和后庭，将公寓楼与街道或中心庭院连接起来。从街道通过主入口进入公共庭院、再进入中心庭院，这种层层递进的过程很独特也很有必要。带有花园的住宅院落为人们提供了聚会和娱乐的场所，对社区间的交流起到了重要的作用。

景观设计的哲学就是创造一个人们承担得起的现代生活环境，将庭院环境和空间的潜力发挥到最佳效果。

景观规划图

1 俯瞰公共庭院

2 中心院落

Located on the Victoria Park Stage two site in Green Square, Zetland, on a 6265m² site, this project consists of 119 apartments in four separate buildings arranged around two large common courtyards. The design for the central communal courts is a complimentary response to the overall Victoria Park open space strategy. The location of NOVA between the Central Park and Tote Park ensures that the larger active space requirements of the projects' residents are met. The NOVA communal courts have been developed with the aim of satisfying the more intimately

scaled "recreational" needs of the residents. The two courts are characterised by bold contemporary patterning, which can be read at ground level or from the overlooking balconies. The communal courts are arranged as a graphic sequence of "rooms", which segment the central open space in a linear manner.

Mcgregor+partners won the competition for the Landcom Victoria Park site with architects, Turner and Associates and the developer Waltcorp. Having had involvement in the project from masterplan conception to final site design has

allowed the delivery of innovative sustainable initiatives by the team in many areas.

The courtyards for NOVA were designed as "pairs of gardens" and are characterised by slender Eucalypts in lawn mounds and complimentary waves of grasses and gravel paths. A grass and stone garden were created to reference the pre-existing ecology and the geology of the site. They are intended to be modern evolutions of the Sydney Bush School of design underpinned by principles of endemic biodiversity. The oval mounds are designed to

create maximum soil depth for tree growth and to allow roots to spread. The courtyards were designed with an internal circulation system with the gardens/public spaces separate from the housing to make a physical separation.

The landscape design of these courtyards responds directly to the environmental design principles of the project to create a healthy living environment, which does not merely focus on picturesque or visual concerns. Our overriding objective was to deliver a project that integrated ecologically sustainable design throughout all facets of the buildings and landscape within a tightly budgeted, construction oriented, delivery process. Of key importance were the spatial and planting aspects of the design, with a planting palette consisting almost entirely of Australian natives. The south sides of the buildings are planted with tree ferns, palms and shade tolerant native species. As Nova has matured it has become a haven for the local fauna, and some of the bird species found here include Finches, Blue Wren and Currawong. Recycled materials with low embodied energy have been utilised such as

1　光与影
2　石头园
3　草木园

229

Sydney sandstone and timber.

The central court is flanked by individual patio courts. Depending on the unit configuration a front and rear court is provided to each unit, which links either to the street or the central courtyard. These internal courtyard entries are designed with a sequence of turns and small level changes to create a varied and tight scale entry to each unit through the garden. The hierarchal journey from the street through the main gates, into the communal courtyard, and then into the garden courts was also important. The individual patio courts are important to the social fabric of the community as it provides meeting spaces and areas for passive personal recreation.

The landscape design philosophy was founded on a desire by our team to create an affordable, contemporary living environment that maximised the environmental and spatial potential of the courtyard for residents.

1　隆起的草坪
2　院落的边缘
3　园内的木板路

集体享有的私密花园 —— 信德上城

A Public Owned Secret Garden — Xinde Residential Community

撰文：安庾心　　图片提供：AECOM Planning Design

总平面图

在广州，过去大多引进东南亚风格的植栽设计，这种比较单一的设计使得广州慢慢缺乏了自己的住宅风格，同时更多中国传统的园林设计风格正在缺失，这是当下最需要解决的问题，作为设计师应该去寻求更多元的适合广州风格的花园设计方式。

该项目位于顺德区容奇大道，是高密度住宅小区，楼盘错落有致，不在同一轴线上，间距也比较近，因此，形成了各种不同的围合空间。在如此复杂的空间尺度上，景观设计师结合中国南方植栽的品种及功能，创造出一系列具有当地特色而且设计新颖、精致细腻的开放式围合花园空间。每个不同的小空间结合各个建筑单体巧妙自然地形成浑然一体的感受。

为了让围合花园空间具备舒适性和亲和力，设计师在整体空间的规划上，尽量将人的行动轨迹梳理清晰，并根据每个空间的不同功能分别进行设计，形成众多不同层次的花园空间。如在比较小的范围内，种植一些植物，安置一张坐椅，形成私密的围合空间，人们可以在其中看书休憩。而对于消防通道的设计，则致力于将其变成广场空间，使其在不缺失功能的前提下与整个环境融为一体，和谐统一。

设计师在围合空间的营造中，同样注重植栽的运用。重点考量植栽品种，多采用现代简洁的线条，打破原来单一的背景，运用错落的植栽营造出活泼而富有层次变化的花园式小空间。植栽的次序变化极大地丰富了园林景观的空间构成，也为人们提供了可选择的空间类型。

设计从小区入口便开始营造出花园般感受，运用花池、树池及错落搭配的台阶，形成亲切幽静的入口。并采用三种颜色的材料，设计出尺度夸大的抽象的现代风格干挂墙体。墙体的围合部分构成开放空间，树与开放性墙体相结合，随着树的生长，枝叶透过围墙伸展出来，形成内外空间的互动。

进入大门首先映入眼帘的是三层竖向跌水的大树池，借由这种围合感，立面插花式种植大树，借用盆景的概念，主景树与配景树形成三个层次，树与三层跌水相映成趣。打开水时体现出水景墙的概念；关掉水后又具备花池功能。树池为人们带来地标式花园入

1 树阵下的自然汀步
2 现代简洁的植栽手法
3 渐变的景观层次

口的感受，并起到影壁作用，转进去才可以观察到后面的景观。

大树池后是另一面水景墙，也是游泳池的一部分。28m长的游泳池，相对地面整体抬高1.5m，需要借助台阶才能抵达，免去了设置围墙的工作，在保证私密性的同时，也顾及到安全性和美观。在游泳池里，由上往下可以看到大树池和三层跌水，营造出宁静的度假感受。山体公园位于游泳池后面，采用森林式植栽方式，山水搭配，加强了游泳池的私密性，减少外界干扰，形成静谧放松的自然空间体验。山体公园延续出一片草坪和林阴大道。树阵广场形成另一种围合空间，迷宫式植栽的运用，使花园的空间感与舒适感并存。

细节彰显用心，细致雕琢的地面铺装、排水盖等，营造出细腻质感的空间。雕塑及各种小品的运用，丰富了花园空间的景观体验。

景观离不开人，它装载着人类的活动与情感。整个大的景观空间里有很多不同尺度及功能的小空间，用来满足人们的不同活动。每个花园空间或由建筑体围合，或由景墙和不同层次的灌木围合。融入生态概念的同时，利用合理的规划，适当搭配具有本土特色的植栽及材料，营造出人与自然充分接近的休憩生活空间，让人们尽情享受花园式的景观空间，获得重返自然的身心感受。

设计回归人们的基本需求，如今的小区花园有一定展示效果，但还应满足回归自然的心灵需求，植栽设计打破传统审美观念，刻意寻求一种全新手法。在广东地区花园的设计多采用东南亚风格，而植栽的设计则千变万化，该项目的植栽设计完全脱离了东南亚风格，突出色彩与质感，在亚热带地区实现四季变化。充满现代感的设计方案主要采用线性的植栽设计手法，还有间列的排序，使灌木在其中发生层次变化。

四季效果在植栽中的运用是种植落叶小乔木，同时加入不同季节开花的品种。在大空间中利用植栽区划分出灵活的小空间，形成站与坐的视觉变化；根据自然变化的光影关系，营造出视觉艺术氛围，使得植栽在不同光影下形成视觉变化；水景墙在阳光下绽放出无限的生命力，流水声带来自然界的灵气。四季的变换、光影和流水形成幽径交错的花园。

花园概念早期来自于个人空间，属于富有阶层享有的特权，但在中国当下则是一种集体共享的公共空间。随着物质文化需求的增长，人们把这种公共空间看成生活中重要的部分，在不同的时间和季节经过这个空间时，会有不同的感受，对景观也将产生不同的细节记忆，从而达到公共空间的个人化。

In the southern region of China, majority of residential landscape design are in favor of South East Asian style, particularly in planting design. This trend has made a good first impression for majority of the people and it becomes the benchmark for the future development. However, this common acceptance of the style also becomes a limitation for further investigation and exploration of residential planting.

Xinde residential community is located in city of Shunde. It is a high-density residential development with a complex layout which resulted in multiple left over outdoor spaces. Landscape designer takes on the challenge and turns these left over spaces to series of innovative, elegantly detailed courtyards and gardens. Landscape designer maximizes the use of diverse planting selection at the south, and creates gardens that are rich in colors and textures.

The diversity of the gardens comes in many forms. Despite the richness of its planting selections, it is also diverse in its scale, proportion and function. A clear pedestrian circulation path shared with the emergency vehicle access connects to all courtyards and gardens sets up a strong overall framework of the design. Each courtyards and gardens are subtly differentiated by the planting palettes, types of outdoor furniture and placement of lighting features.

In Xinde residential design, we are searching for a new planting typology that will go beyond the tropical style. What we are looking for is a garden that still retains its seasonal quality and is truly native to the southern

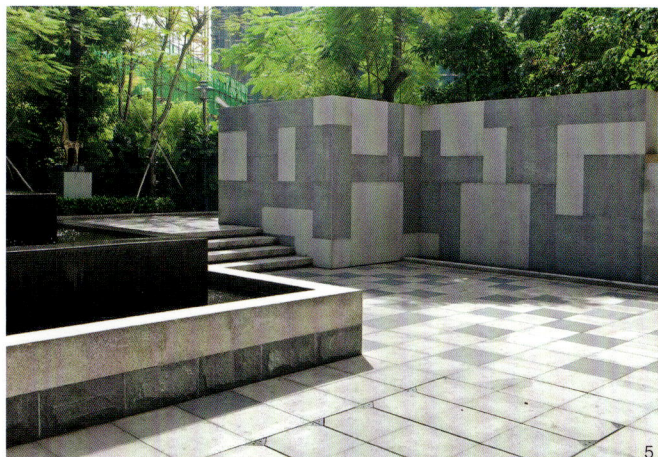

region of China. Planting design is the key driver for Xinde residential community. Using the linear patterns and geometry for shrubs and trees layers, the gardens seems contemporary with a twist in its richness of colors. Each garden has multiple functions and its underlying programs. Some places are intended for single or double occupancy where others are used for larger gatherings and activities. Within the larger courtyard, there are smaller intimate spaces which are separated by tall hedges. It is intended to create private spaces used for passive activities such as reading and meditation within the larger public area. The emergency vehicle access also has a double function; it acts as pedestrian path, as well as it becomes part of the plaza. By merging EVA with the courtyards and plaza, it dissolves seamlessly with the overall design.

The use of semi-open walls, water fountains, sculptural element, and lush vegetation, a hint has been given to the public that one would expect to see a secret garden behind the wall and gateway. The entry is marks by the iconic planter, which creates a strong backdrop for the entry courtyard. The planter plays a double function, it contains multiple layers of trees and shrubs, and it also acts as a water feature wall. Behind this 3 meter tall planter, there is another water feature wall, which is part of the edgeless swimming pool. The 28 meter long swimming pool is elevated 1.5 meter above the finish floor level. By elevating the pool, it will not only avoid building a safety fence, but also gives the pool more privacy.

Landscape designer uses linear patterns for the design of the garden, while mixing plants with different heights, leave textures, and colors. Pending on the day of the year, the garden has different appearances because of the light and shades. The splashing noise of the water fountain, the large terracotta pots, and the wooden furniture, adds an additional cozy ambience to the gardens and courtyards.

People enjoy strolling in a garden, especially after the rain, when sun comes out, the fresh mixed smell of the flowers, leaves and soil reminds us our deepest desire for nature. We all enjoy looking at nature in its simplest form, and without hesitation, we would run bare foot on the open lawn, getting sun bath while enjoying the peace that brought to us by Mother Nature. It is the landscape designer's job to make secret gardens and courtyards that hold the heart for its community.

景观设计 LANDSCAPE DESIGN
景观－设计－艺术

杂志 MAGAZINE

《景观设计》杂志自创刊以来一直致力于发掘景观建设中存在的各种环境问题和设计新潮，为中国的城市景观设计、环境规划和城市建设等提供专业化指导，共同推进中国景观设计行业的发展；《景观设计》以其敏锐的视角和专业的办刊风格而成为同类媒体中的先锋杂志，备受业内人士、政府及开发商的推崇。

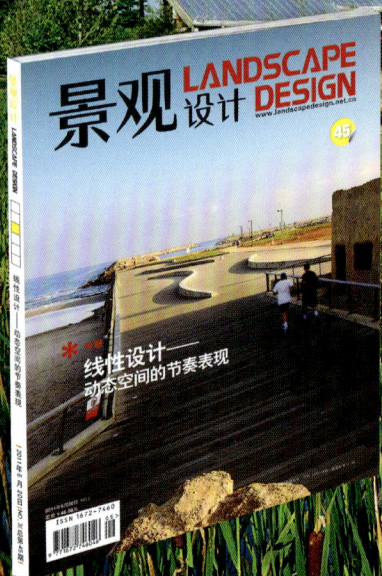

立足本土 放眼世界
Focusing on the Local Keeping in View the World

征订回执单
SUBSCRIPTION

2011年合订本定价288元，共订阅（   ）套
2012年全年共计288元（每期定价48元、全年6期），共订阅（   ）套
备　注：请详细填写以下内容并传真或寄回，以便款到后开具发票和邮寄杂志（此订单复印有效）。

订阅类别：□ 个人　　□ 单位　　　　　　汇款方式：□ 邮局汇款

姓　　名：　　　　　　　　　　　　　　　户　　名：大连理工大学出版社有限公司

电　　话：　　　　　　　　　　　　　　　地　　址：辽宁省大连市高新技术产业园区软件园路80号理工科技园B座802室

传　　真：　　　　　　　　　　　　　　　邮　　编：116023　联系人：宋鑫

单位名称：　　　　　　　　　　　　　　　□ 银行汇款

地　　址：　　　　　　　　　　　　　　　银　　行：大连农业银行栾金支行　　账　　号：34 0400 0104 0000 697

邮　　编：　　　　　　　　　　　　　　　需要发票：□ 是（收到您的汇款后，发票将随杂志一并寄出）

合计金额（大写）：　　　　　　　　　　　　　　　　　□ 否

联系人：宋鑫　　传真：0411-8470 1466

UNITED STATES
On the Road
2011 美国

西海岸景观、建筑项目考察

- **考察时间**
  2011年10月27日~11月6日

- **主 办 方：**《景观设计》杂志社（www.landscapedesign.net.cn）
- **支持媒体：**《城市·设计·演变》《城市建筑》《中国园林》
  （排名按拼音首字母排序）

- **支持网站：** 园林在线、中国人居环境网

- 主办方保留本次考察行程及项目的最终解释权

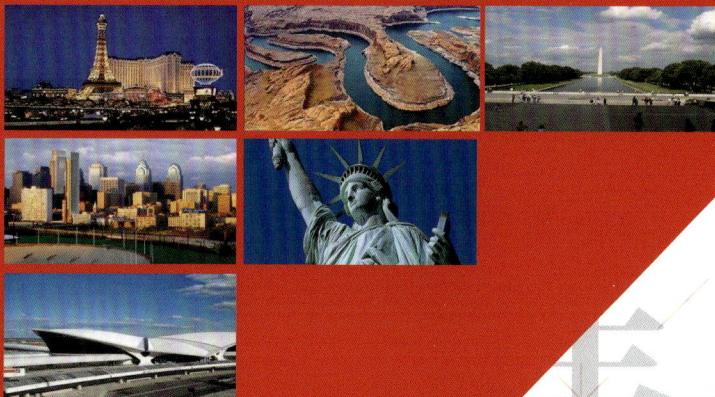

2011年第八届《景观设计》理事会议
Unique Landscape Design from 8 Continents

主题展览 & 主题论坛

城市·设计·演变

展览时间：2011年10月24~26日

地　　点：北京798桥艺术空间(北京市朝阳区酒仙桥路4号798艺术区D09-1)

主办单位：大连理工大学(出版社、建筑与艺术学院)

媒体支持：《城市·环境·设计》《城市建筑》《景观设计》《建筑细部》《建筑与文化》
《设计新潮》亚洲人居环境协会《中国建筑装饰装修》《中国园林》
《中外建筑》（排名按拼音首字母排序）

网络支持：园林在线、中国人居环境网

详情请洽：0411—8470 9075/35